Challenging Assumptions in Ophthalmic Nursing:
a patient centred approach

ISBN 978-0-9556890-0-0

Copyrights Reserved Ryman-Liggins 2008

Challenging Assumptions in Ophthalmic Nursing:
a patient centred approach

Thank God it's finished!

We would like to say a big thank you to our respective families, friends and colleagues who helped us along the way! We couldn't have done this without you!

Grateful and relieved!

Annette & Kim

Annette Ryman MSc, BA (Hons), ENB 346, RGN
Kim Liggins MSc, BSc, SRN, OND, Cert Ed (FE), Cert Mgt (Open) LPC

Website Disclaimer

The websites indicated in this book were correct at the time this book was published and it would be unreasonable to hold us liable hereafter, in respect of these websites and the information on therein.

Whilst we endeavoured to ensure that the information on these websites was correct, following the publication of this book, we do not warrant their completeness or accuracy; nor do we commit to ensuring that these websites remain available or that the material on these websites is kept up-to-date.

Challenging Assumptions in Ophthalmic Nursing:
a patient centred approach

Contents

- Introduction – Why bother?

- The Typical Ophthalmic Patient?

- Exploring Visual Impairment

- Assumptions in the Assessment Process?

- Eye Health Advice?

- Patient Compliance

- Evidence Based Practice and Research

- The Role of the Specialist Nurse – Pushing the Boundaries?

- Professional Issues

- Policies, Procedures, Standards and Audit

- Nursing Models in Ophthalmic Nursing Practice

- Conclusion – Is this it?

Challenging Assumptions in Ophthalmic Nursing:
a patient centred approach

Introduction

Have you ever been visually impaired, even if it was for a few hours due an eyelash in the eye? If not, try it! Occlude an eye for a period of time using an eye patch, or try to obtain some simulation spectacles. This will by no means truly represent what a person with a visual impairment may experience, but it may give you an idea of how it feels to be visually restricted for a short period of time. It's not a comfortable place to be is it? Just remember that you were only trying it. You can take the simulation spectacles, patch or blindfold off but your patients/clients cannot. Yet as nurses, healthcare professionals or caregivers we often forget how limiting a visual impairment can be. We cannot see through the eyes of the visually impaired person. To the rest of us they often look fine. This makes it all the more difficult to identify with the isolation, loss of independence, frustration and devastation that a visual impairment can bring when it first happens and even as it progresses. We often don't appreciate the affect that it has on a patient's friends and loved ones.

Whilst teaching the Ophthalmic Course, we would ask our students to think about and question practice, which is of course what we should all be doing. We were extremely conscious that our students felt awkward doing this and often our efforts to question practice appeared to stay within the classroom environment. It is evident that linking theory with practice will take a little more than classroom discussions and assignments. If this balance is not addressed then we feel that patient care may suffer as a result.

Our passion is that patient care continually improves, particularly in this specialist area. People with visual impairments present in all areas of our hospitals, nursing homes, schools, general practice surgeries and community care specialities. We cannot ignore the fact that visual impairment is a growing issue. It is unfortunately one of those areas that is often neglected, because it is usually only visible to the person experiencing it. This means that making assumptions about these individual's ability to see occurs readily, and is often frustrating to both the patient and the carer. This is really an area that is rarely addressed in literature especially by members of the nursing profession.

We used to say that we couldn't be the only two nurses on this planet who felt this way. We decided to do a literature search and were amazed to find that although there were some papers looking at various aspects of nursing practice, there was nothing that truly reflected how we felt. We found a few rather informative, medically orientated, ophthalmic nursing books, but nothing that encompassed the conceptual discussions, addressing some of the issues, which were close to our hearts. We decided to embark on writing this book. We have tried to tackle some controversial issues and our comments may raise eyebrows, provoke anger, engage you and perhaps make you stop and think about what you are doing, how and why you are doing it. We do not profess to be authorities on ophthalmic nursing but our theorising is based on many years of experience.

We, as nurses are change agents, we influence care, educate and advise patients and their significant others, and what we say and do makes a difference to every care episode we participate in. We are a powerful body when we stand together to challenge or change practice. People with visual impairments are vulnerable, albeit temporarily or permanently, and we need to be strong advocates for good practice and excellent care. If you are not a nurse but are caring for someone who has a visual impairment, we hope that this inspires you to enquire more about the service that your client or loved one is receiving.

This book includes scenarios, real episodes of care, practice - some of which we are proud to share, other episodes, which might make your toes curl. We aim to engage you in debate, to share practice, to get you think about your own practice, and provoke thinking specifically about the visually impaired patient as a whole person with their own unique set of circumstances. We know that this book may be contentious, but we are satisfied that we are attempting to move practice forward.

Join our forum on www.ophthalmic-nursing.co.uk and let us know what you think!

Challenging Assumptions in Ophthalmic Nursing:
a patient centred approach

The Typical Ophthalmic Patient?

Is there really a typical ophthalmic patient? We know, as health care professionals, that patients come from all walks of life, of any age, race or disposition (Stollery et al; 2005). Therefore since the answer to our question is clearly no, why bother writing this chapter? The one thing that each patient has in common is some form of visual impairment or the threat of deteriorating vision, which to the patient can vary from being mildly annoying to devastating. This chapter, therefore, aims to challenge our perceptions of the people we care for who have visual impairments and the potential impact that these visual changes have on their lives as individuals and as part of a wider family group.

One could argue that ophthalmic conditions occur for the following reasons:

- Congenital conditions
- Illness
- Trauma
- The ageing process

Evidence also suggests that smoking may also be a contributory factor in increasing our risk of developing Age Related Macular Degeneration (AMD) by a staggering 50%. Given all of these inherent causes, it means that every individual may potentially visit the eye department at least once in their lifetime. It's a sobering thought, that your vision may be affected at some point in your life and in some cases irreversibly.

What is so special about the ophthalmic patient? This is the part where we need to think carefully about the sense of sight. One could contend that sight does not seem to be too much of an issue until something happens to affect the way that our eyes function. Think about how much we take our senses for granted until we fall ill, have an accident or have a condition which will ultimately affect it. There are a plethora of organisations or groups, which are dedicated to supporting people with eye conditions of which we are ignorant until we or someone we know has a sight problem.

Anyone who has spent any length of time with an eye condition, depending upon the extent of that condition, may convey feelings of fear and anxiety, isolation and helpless, especially in cases where prognosis is uncertain. Vision loss does not only affect the patient but also their significant others; mother, father, siblings, spouses and partners and close friends.

As healthcare professionals, it is important that we always remember that any person presenting with an ophthalmic condition may experience emotional, social, psychological and physical difficulties when faced with reduced vision for any length of time. Imagine a linear scale. On one end we have a patient affected temporarily by a corneal abrasion or foreign body sensation and at the other end we have a patient who has irreversible vision loss due to something like Retinitis Pigmentosa (RP) or Age Related Macular Degeneration (AMD) (visual impairment will be dealt with separately, in greater detail later in the book).

If you are not an ophthalmic nurse perhaps you are thinking what has all this to do with me? In short it has a lot to do with you if you work out in the community, in a general practitioner's (GP) surgery, occupational therapy department or physiotherapy department. Perhaps you work in the trauma field and you are presented with a patient who fell. You ask the patient about their medications, past medical history, recent illness, and mobility restrictions and so on but did you ask them how well they see? Did you check their visual acuity? It might save a lot of time and effort. At least you could rule sight issues out. Many falls are caused by poor vision, perhaps due to cataracts or AMD, or not wearing appropriate spectacles. In 2005, Susan Campbell working for 'Visibility' (formerly the Glasgow and West Scotland Society for the Blind) compiled evidence of this link. This is a really interesting and useful document. If you get a chance we strongly recommend that you read it.

Let's just take some time to discuss the impact of the four suggested causes of visual problems. The aim of this discussion is for the reader to develop a greater degree of empathy with these patients, which will hopefully positively impact on the way that you care for these groups of patients.

Congenital causes

This is not meant to be an A-Z of congenital conditions but it is meant to make you consider what the patient and their carers may be experiencing and how you can adapt your practice to help them. Maybe you are already doing this. Talk about it, share it with other health or community care professionals, we all need this type of encouragement and support. In the meantime indulge us and read on.

Imagine this scenario; a mother noticed a white pupil reflex on a recent photograph taken of her child. She whisks her son off to the GP, who sends her directly to the eye casualty or calls the eye department to make an urgent referral to the paediatric consultant ophthalmologist. The consultant diagnoses a tumour in the eye (Retinoblastoma). Let us put some meat on the bones of this scenario. Mum had to take the day off work, and kept her son from school to bring him to the department. She is anxious about her other two other children and about who will be there for them when they get home from school. Her husband's job is demanding and he often finds it difficult to get away from work early. She has just been told

Challenging Assumptions in Ophthalmic Nursing:
a patient centred approach

that her son has cancer in his eye. What a bombshell! Mum presents in the eye casualty in a frantic state and she hisses at all who speak to her. The child has endured dilating eye drops and a full ophthalmic assessment.

Thinking about this situation from the nursing perspective, how, as a nurse, do you help them through that? What are the implications? Imagine that this was your little boy. You have just been told that your son has a large tumour in his eye, which has affected his sight, and not only that but he needs to commence chemotherapy and have his eye removed (enucleated) as soon as possible. Hopefully your department is well supported with a sensitive paediatric ophthalmic consultant who is equally well supported by the Sensory Impairment Team and ideally there would be someone to sit with Mum to offer her further emotional support if she needs it. As the nurse or health care professional you will know that the main aim of the treatment is to preserve the child's life. The second aim will be to preserve vision (because of the increased risk of the other eye becoming affected) and thirdly to minimise any complications or side effects of the ensuing chemotherapy.

We know what our objectives are, but in the midst of this there is possibly a confused child who really does not understand the full implications of what is happening. Let's call this little boy Paul. He knows that it must be bad because his mother looks worried and upset. He's had drops put in his eyes which stung and bright lights shone in his eyes whilst his head was on some contraption that looks like a machine for torturing people! But everyone is being very nice to him and telling him how good and brave he is being.

If you were his mother how would you feel? In my experience most people in this situation hear "cancer" and "loss", perhaps even blindness even though at present it's only one eye that is affected. It's not that they don't want to hear anything else but nothing else seems to reach them. It appears to be part of who we are, to embrace the limitations of such things at the beginning. You know that they will get through this, but you are also cognisant of the fact that they are entering a grieving process. In my experience, as a specialist nurse, support can be rejected at the point of diagnosis. Mum may decline your support but you have to leave the door open (I usually give a card or booklet with my name and contact details on and sometimes a friendly follow up phone call). In a short space of time, mother and son have had to come to terms with so many different things

- Cancer
- Sight loss – was it Mum's or Dad's fault if it was genetic? Many parents blame themselves
- Need for surgery –being hospitalised and needing to arrange care for her other children
- Need for chemotherapy

Challenging Assumptions in Ophthalmic Nursing:
a patient centred approach

possibly creating a very vulnerable individual. Our Mr Smith is not going to fit neatly into our planned care, because no one truly ever does. Often, as Kim will later argue, the models of care that we purport to use are rarely used to plan care in the outpatients department. We need to identify what he needs and work with him and his family to achieve those objectives. Experience has shown me that how an individual deals with their change in vision is dependent on the general disposition of that person, the level of vision they have and which part of their vision is most affected.

Let us take a step back and use some anecdotal evidence to support the point I am trying to make. An integral part of the retinal specialist nurse's role is to ensure that patients have a link between the consultant and external support services. I once cared for a patient who was slowly losing his sight as a result of diabetic retinopathy (damage to the back of the eyes due to diabetes), he was in his early forties, and was devastated at his loss. His wife went everywhere with him because he was too afraid to go out on his own. He refused a symbol cane, along with any other support from the Sensory Impairment Team, expecting his wife to give up her job to take care of him. The strain on them was telling. Talking to his wife, it was clear that she was frustrated about her husband's sight loss. She confided that her husband refused to tell his friends that he was losing sight because he was too proud. He lost his job and his wife went to working part time to care for him. I spent hours talking with them both, trying to organise support that they ultimately refused, after agreeing to it initially. I listened to their frustrations and hopes of vision restoration. I explained that his vision was unlikely to be restored, but that there may be some improvement, but still they hoped and were ultimately disappointed. This man lost a huge part of his central vision and much of his peripheral vision was affected. He underwent surgery to remove the blood that had leaked into the jelly of both eyes (vitrectomies), which helped to clear the vision to a point. He had numerous laser sessions to try to seal off the leaky blood vessels at the back of his eyes, but to no avail. He continued to struggle to keep his blood sugar levels stable and his vision deteriorated to 6/60 in one eye and 3/60 in the other. At this stage he was registrable as sight impaired. People do not always fit neatly into the way we believe the system should intervene. We can only *offer* the benefit of our knowledge, expertise and support. Patient choice is instrumental in the whole process and accepting their decision is sometimes difficult when our support is declined.

Trauma

Probably one of the most common problems we see in the Eye Casualty is the patient attending with a corneal abrasion or foreign body. For this I will use anecdotal evidence as an example. My Mum is a very young sixty. She wears spectacles for myopia (short-sightedness). She lives alone and manages poorly without her spectacles. She was walking along a pavement where there were

Challenging Assumptions in Ophthalmic Nursing:
a patient centred approach

some low overhanging branches and as she was passing them, a branch brushed her face. Unfortunately the branch caught her in her better eye. She said that her eye became painful and started watering. Eventually she could stand it no more so she attended the local accident and emergency department. After waiting for more than three hours without any analgesia she was examined and she was identified as having a corneal abrasion. Her report stated that Mum found it difficult to keep her eye open and that she complained a lot. It is of course difficult to stay objective when discussing someone so close, but I know that Mum has a high pain threshold so I found the report rather upsetting. Just imagine if it was your relative?

The nurse who attended Mum clearly had no idea about the degree of pain that a person with a corneal abrasion can have. No advice was given to her about taking oral analgesia, a single patch was placed over her eye and the nurse sent her home with a tube of eye ointment (Chloramphenicol). Unfortunately no one had thought about how she would see when she left the department. Her spectacles did not fit over the patch, which she left in place because she was worried about removing it. No one had asked how she was going to get home and how she would manage putting the ointment in. I spent two days and nights with Mum cooking for her and applying the ointment as prescribed and giving her regular analgesia. She couldn't see anything until the third day when the eye had settled.

Some simple considerations would have made life much easier. She did not feel as though she had been treated as an individual and was clearly not. Her episode of care not only impacted on her and her ability to manage, but on the rest of our family.

Age Related Problems

Age Related Macular Degeneration (AMD) is a leading cause of central vision loss in the over 60 age range, of which there are two types, wet and dry. It's soapbox time! This is not just "wear and tear" at the back of the eye, this is a devastating eye condition to many individuals, who feel that they might as well not be alive (especially at first point of diagnosis), once their central vision degenerates completely. Unfortunately all too often this is just another condition to add to other medical conditions that the patient may have. It can lead the patient to feel despondent and depressed. No one wants to lose their independence and many patients feel as though they become a burden to family and friends once they lose this crucial part of their vision. Thankfully many patients adapt to their vision with great patience and perseverance on their part and with much love and support on the part of their significant others.

I once encountered a delightful man, who I shall call Mr Smith. He was 77 years old and lived with his "disabled" wife. He had been diagnosed with bilateral exudative (wet) AMD. His vision had significantly deteriorated over a period of a couple of weeks.

Challenging Assumptions in Ophthalmic Nursing:
a patient centred approach

Speaking to Mr Smith his main concern was getting back home to his wife after a very long one-stop clinic. Despite my best efforts to offer him support in terms of the low vision service and social services, his greatest concern was to be able to continue to care for his wife. He was responsible for administering all medications. At that point in time he was still able to read the medicine bottles, but my concern was what would happen when that became a problem. I gave him my details hoping that he would call should he start to struggle with reading, but he seemed disinterested. The frustration I felt at not being able to help this man was immense but this is an important point here. The care was what I wanted to give not what the patient felt he needed. Standing back was really difficult. There are no easy solutions when dealing with visually impaired patients. Each patient deals with their loss in different ways and often it's not the way that we would choose for them.

If that all seems really logical then clearly you have will not have felt challenged. Using the scenarios I have attempted to show that anyone may experience some vision loss at any given time in their life. This is not an exclusive condition and can prove more than inconvenient. Vision loss not only affects one individual, it affects a whole group of people. People with a visual impairment deserve and require special consideration. Encountering the person who has a visual impairment, no matter how temporary, requires a great level of empathy and understanding. We are always taught to treat others as we would like to be treated. When you have never experienced a visual impairment it is probably difficult to imagine the type of consideration that one should proffer but we all know how to treat people with dignity and respect, part of which is talking to the patient and enquiring what their needs are. I will take this discussion further in "Exploring Visual Impairment".

References

- Stollery R, Shaw M and Lee A (2005) Ophthalmic Nursing – Third Edition – Blackwell Publishing
- Campbell Susan (2005) Deteriorating vision, falls and old people: the links for Visibility (formerly Glasgow & West Scotland Society for the Blind) – http://www.visibility.org.uk/what-we-do/research/Falls-Report.pdf

Useful links

- www.rnib.org.uk
- RNIB Helpline: 0845 766 9999
- http://www.health.qld.gov.au/fallsprevention/brochures/Vision.pdf
- http://www.optometrists.asn.au/ceo/backissues/vol88/no4/4731
- http://www.patient.co.uk/showdoc/27000750/

Copyrights Reserved – Annette Ryman & Kim Liggins 2008

Challenging Assumptions in Ophthalmic Nursing:
a patient centred approach

Exploring Visual Impairment

In the first chapter we challenged possible perceptions of the typical ophthalmic patient and in doing so touched on the implications of experiencing a visual impairment. In this chapter there we will attempt to challenge the reader to think about and correlate the subjectivity of vision loss with the quantifiable actuality of vision loss. We will be looking at the importance of visual acuity testing and how it is performed. This chapter will further endeavour to challenge our perceptions of vision loss in terms of both temporary and permanent sight loss and the way in which we manage our patients in both situations. For those readers who are not nurses, this will hopefully broaden your view of your client's or significant other's clinical experience and add to your understanding of how best you can support your client.

Visual Acuity testing

One could argue that the visual acuity assessment is probably the most important assessment used for a patient. This chapter does not address "how to" test visual acuity, it is more about challenging the process and our attitudes towards performing the assessment. In my view, visual acuity testing begins when you call your patient. Why? Because watching your patient navigate towards you can give you some useful information about how they are managing. Unfortunately, in my experience, nurses and doctors often call a patient and disappear back into the treatment or examination room, leaving the patient wondering where the voice came from. Taking time to meet your patient can prove extremely useful and often makes the whole visual acuity testing process easier for both the patient and the nurse.

Once you have your patient seated comfortably then you begin. Let me just ask you a few questions to reflect on. Do you think that you are complacent about performing the visual acuity test? Do you always watch the patient when you are doing the assessment? Do you note what part of the vision the patient is using? Do you record that the patient moves their head when checking their vision? Do you even notice what the patient is doing or are you too busy writing or looking at the chart? What do you do when you notice that the patient's vision is worse? Do you ask the patient how they are managing? Do you notify the specialist nurse or query need for registration or low vision assessment?

Performing a visual acuity assessment at first seemed easy to me. I was first taught by a support worker who told me that if the vision is worse than 6/12 then use a pinhole to see if it improves. At the time I was an agency nurse and just needed to stay in a vision bay and "do visions". I noticed that some of the patients claimed they could see less with a pinhole and being the enquiring sort; I went away and started looking at books to do with the assessment. I knew that the 6 at the top of the equation meant 6 metres but I wasn't sure about the 12 at the time. I felt as though I needed to know why I was doing it and what it all meant. When I first learnt how to do a visual acuity, I was so focused on listening to the patient, read out the letters or numbers and watching the chart that I paid very little attention to what the patient was doing. Of course then I had to make sure that I recorded the vision correctly. No one had told me that I needed to watch the patient.

The patient always wants to do well, and in my experience will always utilise whatever part of their vision that they can, even down to cheating which is rarely their intention. They will turn their heads to the side so they are looking out of the corner of their occluded eye. They may look through their fingers if they are not using a proper occluder or the palm of their hand to cover the eye. As an ophthalmic nurse this may seem like "teaching grandma to suck eggs", but at times we forget or fall foul to a variety of distractions. One nurse I know managed to get 6/6 out of an artificial eye and the patient said absolutely nothing to her. It was the consultant who saw the patient who finally picked it up. Admittedly Log Mar charts have probably compounded the problem causing us to be more distracted by the new letter formation and the different scoring method. However we still need to be cognisant of how the patient is reading the chart.

Visual acuity testing is a quantitative assessment, which makes it influential in deciding what happens in terms of referrals, where vision has deteriorated. Clearly most ophthalmologists will question a visual acuity that does not marry with their findings, but it is important to get it right from the outset. As ophthalmic professionals we look at the numeric data and make certain assumptions based upon this assessment. We determine whether a patient is registrable as partially or severely sight impaired and provide or deny treatments in respect of it. All the more reason why we should be paying close attention to teaching staff about obtaining an accurate visual acuity and about the implications of altered vision.

If you are responsible for measuring visual acuities, think about the way that you were taught to do it. When I did my ophthalmic course, Kim used to talk to us about Martians. Why? Because she wanted to get us used to explaining what we do and why we do it in such a way that anyone could understand. Would you feel totally confident about teaching visual acuity assessment to one of Kim's Martians? Could you give full explanations about the measurements you are observing and why the patient can see better or worse with a pinhole? If you are a trained ophthalmic nurse I would hope that this is the case. However, there are other services such as the diabetic screening services and accident and emergency departments who may not feel as confident in the theoretical basis of visual acuity testing. I lecture on trauma courses and very few staff are aware of

Challenging Assumptions in Ophthalmic Nursing:
a patient centred approach

the basics of visual acuity assessing, even down to asking whether the patient wears spectacles or even using a pinhole to see if the vision improves. Should we be concerned? Should we not be supporting primary care givers in this area to ensure that patients are referred appropriately and expeditiously?

Let us imagine that we have done our visual acuity assessment and we notice that the patient's vision has deteriorated. The simplest and most obvious question would probably be is the patient wearing the correct spectacles? As I am sure you are aware, some people have several pairs and there is usually a story behind each pair. Do you ask the patient how they are managing and if they have noticed a change? Or do you go ahead and put the notes out ready for the doctor? Do you ask the patient whether they feel that they would like some social support? Do you think about making a referral for visual impairment (RVI)? Do you ask whether they are struggling with reading and seeing detail and discuss referring them for a Low Vision Assessment? Perhaps you might think of discussing your concerns with the doctor or specialist nurse (if you have one). Do you know what constitutes low vision or registrable vision? Do you know when to advise your patient to stop driving?

I once saw a patient who had 6/24 in both eyes and was still driving. I asked the patient when he last attended his optometrist and he said that it was 6 months ago. His optometrist had referred him into the outpatients, but according to the patient he had not been told to stop driving. He said that he felt that he could see perfectly well and that he only drove short distances anyway and never in the dark. He was devastated at my telling him to stop driving and that he needed to notify the DVLA about the deterioration in his vision. This was his independence. He had not considered the danger that he posed to both himself and others, even on short journeys. My feeling was that he knew that his vision was worsening but he did not want to accept it. He felt that he was not doing any harm by driving "short distances and never at night". This is by no means an isolated case. I am sure that you have anecdotal evidence to support this scenario. Just think for a moment, you have been driving for most of your adult life and someone tells you that you can no longer do it. My heart always sinks when I see frustration, despair and disappointment in the faces of patients that I have had to say this to. Some patients take it better than others do, but nevertheless it seems like one more nail in the coffin of independence to the patient.

Let us look at a definition of 'Visual impairment'. It could be described as a consequence of a functional loss of vision, rather than the eye disorder itself (NICHCY; 2005). Visual impairments for the purposes of registration are divided into two categories

- Severely Sight Impaired/Blind – "so blind that they cannot do any work for which eyesight is essential."
- Sight Impaired/Partially Sighted – "substantially and

Challenging Assumptions in Ophthalmic Nursing:
a patient centred approach

permanently handicapped by defective vision caused by congenital defect or illness or injury." (CVI 2003)

As previously suggested, an important point to note is that a visual impairment although quantifiable is also subjective. What does that mean? Here's an example, you are presented with a patient who has best corrected visual acuities of RVA 6/12 and LVA 6/9. You were perhaps considering that the patient is still within DVLA standards, so their vision is fine. However, this patient used to have 6/6 vision in both eyes. To that patient their vision has seriously deteriorated. They may not feel quite as safe when they drive, as they used to. What would you advise them to do? Hopefully you would suggest that they do not drive if they feel unsafe despite the fact that they still meet DVLA requirements. The point is that one cannot make assumptions about how well a person sees purely based on measurements. I heard of one patient who loved to do astronomy; he used to have a visual acuity of 6/5 in both eyes and had started to develop cataracts. He had only lost a line of vision but it was creating all sorts of problems with him pursuing his interest. He asked the consultant if he could have the cataracts removed with a vision of 6/6 in both eyes. To most of us that might seem a little drastic but to the patient it was vital that he had the surgery. Think about the patient who has this type of vision and may be experiencing peripheral vision loss, or difficulty with night vision.

What we are talking about here is quality of life. It is also important to note something which may seem somewhat strange, some people, and particularly elderly people, manage perfectly well with visual acuities of 6/36 (Well that's what they will tell you). You will ask them how they manage with cooking, making a cup of tea and performing activities of daily living and they will say that they are fine. Often is the case that they are so used to a poor level of vision and have adapted to it over time, especially with patients who have dry Age Related Macular Degeneration (ARMD), that they do not feel any need for intervention. Perhaps if you asked questions about any recent falls or how they are managing with shopping you might gain a greater understanding of how they are truly managing. It may be that if you visited the house of that person, you may note that it is not as clean, but that they do actually manage. One cannot assume anything. Patients often develop coping strategies, which may not be altogether safe. One gentleman I met had blisters on his fingertips because he had been putting his fingers in the top of his teacup to see when the cup was full. He left the department armed with a liquid level indicator and a referral to the Low Vision Department, which he consented to and was grateful for, but as we know, intervention is not always wanted.

I was rather shocked that most patients will often hide the fact that they cannot see. Admittedly I have noted this predominantly in our younger patients, but arguably anyone may feel this. I do not know

Challenging Assumptions in Ophthalmic Nursing:
a patient centred approach

why I was shocked because my mother is embarrassed that her hearing is quite poor and has been reluctant to share that for many years. People seem to feel embarrassed about something that they have no control over. In my naiveté I always thought that people would want to be helped. I came across a young man, early 30's, who was still driving with very poor vision and he refused to be registered or give up driving because of the stigma attached to being partially sighted. I would suggest that it had something to do with having to face the fact that his vision was deteriorating, were he to be registered. He was very angry and had a number of issues to deal with including his wife's depression and their young child. He was dealing with the problem the best way he knew how and I continued to talk to him and support him as and when he asked for it. There are so many cases I could discuss, but at the end of the day, vision loss is a process and each individual will experience the process in a different way. Kübler-Ross's bereavement process best represents this process because vision loss causes some form of grieving. Often each stage is not clearly identifiable and patients may be experiencing a couple of stages at once or they may even feel trapped in one of the stages. The difficulty for some patients is that they are not just dealing with vision loss. They may be dealing with loss of a loved one, or isolation because family do not live close by to support them, loss of independence which is not familiar to the grieving process, often serving to compound the sense of loss. So we see that this is no longer about numbers or letters on a chart, this is about coping and adapting.

The problem with vision loss is that it is generally invisible to all but the patient, unless the patient has an obvious eye condition. People generally make the assumption that someone has a visual impairment if an individual is carrying a white cane, walking with a guide dog or wearing dark glasses, and even then one cannot be certain with fashions as they are. Patients I have spoken to get very frustrated when out in crowds of people because they feel that people bump into them. I get the feeling that patients feel that others should make allowances for them, but I always explain that no one can see through their eyes and they are not sporting a sign saying "I have a sight problem", unless they are carrying a symbol cane. Speaking to some of my patient group I was made aware that some members have adopted the habit of apologising when they bump into people, even though it might not be their fault. When I asked the group members why, they said that it was easier to apologise than to get annoyed. I think I understand why but it fills me with sadness to think that these people are apologising for their visual dysfunction.

When people appear generally well without any obvious disabilities, I believe that many assumptions are made without realising their limitations. Think about this scenario; you encounter a very angry teenager who has lost sight in one eye and can barely see 6/36 (the top 2 lines on a Snellen chart) in her better eye. She has been told that the vision in her poor eye will never return but that it might

Challenging Assumptions in Ophthalmic Nursing:
a patient centred approach

improve in the other eye. She has been put on huge doses of steroids and is to be commenced on cyclosporin. She has a very supportive family who visit daily. She navigates well around the ward. What assumptions might be made here? She has been aggressive towards staff and is seen to be causing a nuisance. Most of the ward staff do not wish to care for her and generally leave her to her own devices. Her menu cards are left on her beside table and collected without discussion. She struggles to fill them in and accidentally orders food that she doesn't like. How would you deal with this teenager? Staff are clearly not there to take abuse, but one would hope that a nurse would be considerate of her limited vision and her grief. This young lady may be one of several other demanding patients. This patient is going to utilise much of your time and energy and probably test your patience. It just takes a moment to think about the person in your care, but it takes time and effort to support them. These patients and their respective families need so much emotional support. Any ophthalmic-trained nurse would hopefully know that this patient is grieving her vision loss and that in part may result in her behaviour.

I know that I have already raised the following point but its relevance is obvious in this context. We must always be mindful of the fact that we can only *'advise'* the patient and *'offer support'*. This means patient or client choice. The person with the visual impairment is given time to absorb the information given to them and then come to terms with what has been said to them. We cannot force anyone to receive care and support offered and quite rightly so. However, my concern often lies with the knowledge that patients will struggle when they lose their central vision, through AMD, or peripheral vision loss if they do not comply with their drop regime for glaucoma, for example. I am by no means advocating paternalism but I have to admit to feeling frustrated with what we commonly call 'non-compliant' patients.

My area of expertise is working with retinal patients and I felt challenged to make people aware of how we were treating patients with visual impairments. I started by attending a fantastic course run by the RNIB, called the Eye Clinic Liaison Officer course. It really challenged me to think about my practice in general, about service improvement and about setting up a visual impairment awareness session with other ophthalmic nurses. It made me more cognisant of who these patients really are and what they need from our service. We cannot make assumptions. We need to make our services accessible to all and that includes increasing the font size on our patient information leaflets and letters, providing tapes and Braille documents, signage, and making environments conducive to people with visual impairments. I am about to have one of my soap box moments but have you ever tried to get appointment letters to be sent out in a 14pt font or greater? I was surprised at how challenging this was. You would think that this would be so easy. Sometimes I wonder whether we are actually catering to the needs of the patient. We work in a speciality that deals with low vision, where a huge number of patients live alone, especially in my particular speciality. Sometimes I feel as though we are marginalising the very group who needs the most support.

As a result of my passion for raising awareness for the visually impaired I joined forces with Social Services Visual Impairment Team and the Low Vision Services Committee and I co-ordinate and run a Visual Impairment Awareness Session for the Trust. In it we talk about common causes of vision loss, we have someone, who has a visual impairment, to come and talk to the group to share their experiences. We also do some practical sessions such as sighted guide technique and wearing simulation spectacles to give group members some idea of how it feels to have part of their vision taken away for a short time. This is by no means a real indicator of what it is like but it usually creates the desired discomfort and uncertainty that one may experience with a visual impairment. All this session does is to raise awareness and to challenge people's perceptions of what a visual impairment is. It has proved extremely successful but I have to confess that most sessions are filled with Occupational Therapists. I would like to see more of other team members, doctors, nurses, receptionists, secretaries, domestics, in fact any member of the health care team. What are you doing in your area to make staff more aware of people with visual impairments?

References
- Kübler-Ross Elisabeth (1969) On Death and Dying - Macmillan – New York http://changingminds.org/disciplines/change_management/kubler_ross/kubler_ross.htm
- Referral to Social Care - http://www.rnib.org.uk/xpedio/groups/public/documents/publicwebsite/public_registrationnew.hcsp
- Partial/Severe Sight Impaired Registration - http://www.patient.co.uk/showdoc/40000897/
- National Dissemination Centre for Children with Disabilities (NICHCY) http://www.nichcy.org/pubs/factshe/fs13txt.htmp
- Certificate of Vision Impairment (2003) www.rnib.org.uk

Worth a read
- Crossland MD & Culham LE (2000) – Psychological Aspects of Visual Impairment – Optometry in Practice Vol 1 21-26
- Rose K (2000) Vision Testing in the Out Patients Department – Ophthalmic Nursing - December Vol 4 No.3 pp24-26

Challenging Assumptions in Ophthalmic Nursing:
a patient centred approach

Assumptions in the assessment process?

Without realising it we often make a number of assumptions about a person, especially if they look generally fit and well. Assuming that a person can see or not may be one of those assumptions, depending upon how they present. I would argue that more often than not, assumptions in the realms of ophthalmology are based on the numerical information obtained at the point of visual acuity testing, as previously suggested. In this chapter I want to explore the assumptions and conclusions we may find ourselves forming when first encountering a person with a visual impairment. This discussion will be based around a person being assessed in both acute and outpatient settings. It is hoped that through this discussion the reader will reflect on how they can develop assessment strategies to ensure that episodes of care are totally patient focused.

In the acute setting, such as the eye casualty, a person presents with an eye problem, of which the attending nurse attempts to ascertain the nature, duration and severity. Think about where your assessment of the patient begins. I was taught that assessment should begin at the point of calling an individuals name and watching them walk towards me. If you do not do this and just call the patient and disappear into the assessment area I would argue that you have already gone down the road of forming assumptions. Why? You are assuming that firstly the patient heard you, then that they could see you and their surroundings enough to navigate their way to you, and possibly that they are capable at moving at your pace (which usually goes into the realm of light-speed on a busy day). I am sure that you are beginning to see how easily we fall foul of this behaviour. Non-verbal cues and using basic observations may assist the assessor in formulating some notion of how this patient is affected. It is not always the case but it's certainly a good starting point. The initial observations may comprise of noticing that the patient is perhaps wearing dark glasses or something as subtle as a slight head tilt. Individuals with a visual impairment often develop coping strategies and it's always useful to observe and record these to ensure a complete assessment is made.

Having identified some useful information by greeting the patient and taking them through for the history and triage part of their assessment, we assume, or perhaps to be fair, hope, that we are going to be able to glean an effective history from our patient within a short space of time. Time is often a huge constraint on this process, which is often why some things may be missed in the initial assessment stage. In the first chapter we identified that each person with a visual impairment will present with their own unique set of circumstances. They may arrive with a concerned spouse, relatives or friends. They may have been sent there directly from work, school, by their GP or their optometrist. They may not understand the way that you phrase your questions, they may be hearing impaired, have learning difficulties, or there may be a language barrier. To top it all they are visually impaired or visually affected. Given these factors do you feel that you are able to effectively assess all patients who attend your department? Do you take the time to ensure that you have all the relevant details? I have to admit to have felt like a super sleuth on occasions, trying to prise information out of patients who do not willingly volunteer facts about their general health and their eye condition, especially those patients who cannot remember their medications or what they take them for. On some occasions you are presented with a patient who appears to have a few different problems, but they fail to mention the main reason for attending. It can prove one of the most painstaking, frustrating episodes of patient care, for both the patient and the nurse.

Let us examine this part of the assessment process. Before we even take the history we may have already made some assumptions about our patient. For example, when you find out that the patient is hearing impaired and they bring someone with them, do you assume that you need to only address the accompanying person or perhaps that you need to bellow at the patient for them to hear you? It is well known that a woman's voice is much more difficult to hear for someone who has a hearing impairment. Given that most nurses are of the female variety this is something we should be cognisant and considerate of. If the patient is wheelchair bound and accompanied by a carer or family member do you assume that they have no understanding of their condition or would prefer for you to discuss their situation with their companion? I have to say that this situation is something I have seen nurses assume this very notion on a number of occasions. If a child is with their mother or father, do you assume that they have no insight into how they feel? Do you see how easy it is to not consider the affected person? Can you also see how much you can miss or glean depending upon your approach? Sometimes you will find that the patient has not been as aware of their eye condition as other family members or friends, in which case they are probably going to be useful contributors to the assessment process, but clearly assuming anything in such a situation may prove detrimental.

Let's revisit our patient with a visual impairment. Here we know that there is likely to be a high level of anxiety with a patient who presents to the eye casualty. Do we always consider such factors when we perform the initial assessment? There may be a resounding yes in your mind, but at a busy time when things are a little bit hectic; these added extras can become lost in the assessment process and may remain unconsidered. In my experience, anxiety causes many patients to forget the type of medications they are taking for example. They may forget how long they have had their symptoms for or when they first noticed a change in their vision. Your patient may neglect to tell you certain aspects of their eye condition, for example that they have trouble seeing in the dark, unless a specific question is asked. Triage is supposed to be a quick assessment. I was once told that I should spend no more than 5 minutes with each patient, which is totally unrealistic if we are taking into consideration individual patient care. However, it is also unrealistic to spend ½ hour with each patient. One may still argue though, that at times, this part of the process is far too rushed. This often causes human beings to become absent minded, especially our older patient group, leading to a potentially irritable and anxious individual. Do you make allowances for this? Do you take the time to introduce yourself and tell the patient what you will be asking them so that they can gather their thoughts and relax? Do you ensure that you are interviewing a patient where their dignity and privacy will be maintained? At the end of the day common sense should prevail, you treat others as you would want to be treated, but do we have the capacity to do this and if not, why not?

When we ask a patient about their particular problem are we really paying attention to what they are saying? Do we always remember to ask about severity, duration, type of pain, irritation or discomfort? Do you have a pain score and if so do you use it? As nurses, do we have some preconceived notion of how painful or uncomfortable a condition should be? Do you then assess the patient accordingly, without really taking on board the degree of discomfort or pain that the patient is expressing? Do we accept that some patients have a greater threshold for pain than others? For example a patient complains of a foreign body sensation. If that foreign body sensation is causing profuse lacrimation (watery eye) and pain do you still categorise it as a straightforward foreign body sensation or do you express the degree of discomfort. I am aware of the simplicity of these statements but sometimes it's good to go back to basics. Do you always take individual circumstances into consideration? Are you looking at the whole patient when making the assessment, rather than just considering the presenting eye condition? If so, does it make any difference to the care that you offer? Have you noticed how the largest group of patients that you see fall in the middle category of the triage model? This often leads to a back log of patients waiting, who are often in pain, requiring a full eye examination, elderly, diabetic and other high risk groups.

All said and done, I know that an in depth, assumption free, assessment is the ideal and one could argue that it is just not possible. After all, where do we record all the information from the above assessment process? I don't know about where you work, but we have very little space on our casualty cards to write everything but the bare essentials. Perhaps this is something that we need to address? Perhaps a more detailed assessment at the point of triage would mean less time with the doctor? Experientially though, I would contend that this is often not the case. Once the patient has had their initial assessment, with the nurse, they have a tendency to remember other key facts when they see the doctor.

Challenging Assumptions in Ophthalmic Nursing:
a patient centred approach

Janice Ledford has written a book about "Ocular History Taking" (2006) for doctors, which would arguably be useful for Advanced Nurse Practitioners or Clinical Nurse Specialists in the Eye Casualty, or in fact any nurse who regularly takes an ocular history. In my view this book demonstrates how to ask the right questions for presenting symptoms and takes a pragmatic view of history taking. It discusses adapting or modifying language to suit your individual patient. Ledford emphasises the importance of asking broad questions and then narrowing them down to make the questions more condition specific. It embraces a more holistic approach to ocular history taking. At this juncture I hasten to add that we do not have shares or interest in this book, but found it a useful study companion on past ophthalmic nursing courses. The current practice in our unit is that the ophthalmic nurse sees the patient first, triaging them, taking an ocular and medical history from the patient. This begs the question though whether we actually need the initial assessment, since it seems that ophthalmic doctors are taking a more holistic approach to assessment. You may have a raised eyebrow following that last assertion but read on. Your department may have already decided that a patient just needs a quick assessment, quick visual acuity assessment then select questions; are they in pain? Has their vision reduced suddenly? Have they experienced recent trauma? Depending upon the responses the nurse will categorise them as urgent, semi urgent or non-urgent (or allocate them into a colour category – depending on the assessment model used).

Let's examine the idea of triaging. I have observed other departments in other hospitals where a very short initial triage takes place, which seems to work equally well. Do you think that it's worth taking a history from the patient when the doctor is going to ask the same type of questions? It is my experience that we perform, what we think is a thorough assessment of our patient and off they go to the doctor and they tell them something extra or something different. I know that it used to frustrate me. Perhaps one could argue that between the two assessments enough information will be gleaned to assist with diagnosis. To which I hear the cry from management "…is this a cost effective use of the nurses time?" I would argue that taking a comprehensive history will ultimately make for a more effective and expeditious treatment plan. All said and done, since the patient is at the heart of this we certainly need to continue to take a flexible approach to their care. Is there any way that we can be certain that we are always going to get all the information first time around? The assumption is that the patient will tell us everything and at the end of the day and that's all we have to go on until we see their full hospital records. We are clearly not in a position to second guess what patients tell us in a trauma department. All we can do is listen, observe, examine and record our findings as the patient presents to us.

It has been my experience that listening is a key skill in nursing especially in this area. It is also my experience that it is not a skill that

Challenging Assumptions in Ophthalmic Nursing:
a patient centred approach

is as well utilised as it should be. Some nurses start writing the moment that the patient starts talking and they spend more time looking at the chart during a visual acuity assessment, rather than observing what the patient is doing. Often the patient alters or elaborates on their story and in some cases does not get to the crux of the reason they have attended until much farther in the assessment process. Here, the assumption is that the patient is going to get to the point straight away. It's often wise to talk to the patient first without making notes to ensure that relevant information is obtained and recorded.

We have probably well and truly covered the assessment process in an acute setting, but what about the patient attending the outpatient department for their first visit? Much of what has been discussed above is relevant here. Probably the first assumption we would make is that the patient was able to read the letter. This usually gives them the information about dilation drops, and not driving to their appointment. Obviously the patient may miss the appointment if they were unable to read the letter or to get someone to read it for them. This usually initiates the sending of a "Did Not Attend" letter resulting in their discharge. Is that ever picked up? Does someone go through the notes and question and investigate a non-attendance?

Let's go from the point at which the patient arrives. I would argue that we may make the assumption that the patient fully understands why they are attending. On numerous occasions I have asked a patient whether they are clear about why they are attending, with less than favourable responses. Patients have said things like "I was hoping to get some new glasses" or "no one really explained anything, they just said I needed to see a specialist." It has become increasingly evident that the patient does not expect to have done half the things they end up having done. Don't forget that many of my patients are elderly so it is fairly common to hear such things. Rather than making any assumptions here I have assumed the practice of asking the question, "Do you understand why you are here and what is going to happen today? Because I am happy to remind you and go over things before you go back out and take a seat." This usually makes for a more pleasurable (if you can really call it that) episode of care.

I make it a practice to ensure that we send out information booklets about the service and what to expect prior to attending. I generally assume that these have been read and understood. If I see the patient before they attend then I tend to go over the information in the booklet and let them know where to find my contact details. I then assume that not only will they have read the information, but that if they have not understood it, that they or a family member will call me to discuss it. Am I wrong for making such assumptions? Many of my patients come for investigations and look as though they have barely slept all night, despite my providing them with information booklets.

Challenging Assumptions in Ophthalmic Nursing:
a patient centred approach

When I ask how they are feeling, they usually say that they are nervous because they have to have investigations or they are not clear about what is going to happen. The one thing that I have decided is that repeating the information when needed is in the patient's best interest and that's what we are observing as nurses. As frustrating as it can seem at times, we should not assume that just because we have done everything some patients may still require extra information. In our PDT Service we review patients every 12 weeks and for some of those patients I have to explain the same information on each of their respective visits because their short term memory is so poor.

When a patient leaves the department after their consultation, do you then assume that they will know who to contact should they have any concerns? Do you assume that they know how to instil or take any medication that has been prescribed? Do you assume that they understood everything that they were told during the consultation? Making assumptions is a human predilection, but I would argue that sometimes we just have to take step back and make sure that we do not fall foul to assumption making in the assessment process. In doing so we could ensure a positive, balanced and informative episode of care for the patient.

You may think of other assumptions, but I hope that this is food for thought.

References

- Ledford Janice K. (2006) – The Complete Book to Ocular History Taking – EyeWrite Productions – Slack Incorporated

Challenging Assumptions in Ophthalmic Nursing:
a patient centred approach

Eye Health – advice?

I think that most of us agree that prevention is better than cure, but how do we apply that in our roles as ophthalmic nurses? Is health advice something we automatically give to each of our patients? If not, why not? I just want you to think about your reasons for not doing this. There could be a plethora of reasons which may well be outside of your control. Is time a constraint on advice giving? If this is the case, what are you doing to address this issue? Are you reluctant to share advice with your patients because you feel as though you are "teaching grandma how to suck eggs" or do you just not feel confident enough about advice giving? If so, how do you plan to remedy this? As health care professionals, health advice giving should form an integral part of holistic care. I am not just talking about verbal advice here, but any form of information be it in leaflet, CD, tape or if resources allow DVD format, always taking into consideration your patients visual capabilities. Are we reaching out into the community or liaising with practice nurses and general practitioners to discuss eye health? Are we as nurses keeping up to date with current research in relation to eye health? If not, why not? This is meant to challenge you. It's as uncomfortable to write as it is to read, because I too have been guilty at times of not being on the ball as far as research goes. This chapter is meant to raise issues such as these and encourage you to share practice, question practice and be proactive in improving care.

One could argue that health awareness has moved into a wider sphere of interest, because we as a society have changed the way that we live in terms of diet and smoking and our approach to our own health. Obesity is becoming increasingly common and with that a variety of complications which derive from it. We know that a systemic problem such as high cholesterol or hypertension may affect the vision in some way. Stress in conjunction with such conditions can cause all sorts of complications. As we said earlier, prevention is better than cure, so how do you warn patients about these, when they are often indications of long standing problems, which may or may not be dealt with by the patients' own general practitioner? How can we reach these patients? Shouldn't we be looking at discussing eye health before patients attend our casualty with vision loss? If a patient has diabetes, yearly screening is usually set up for them so that their eye health can be closely monitored and a prompt referral made should a potentially sight threatening retinopathy be discovered. Whilst diabetes is rather well covered in this respect, what happens with other eye conditions? Do you have well established eye health connections with your primary care trusts? If so what do you do and are you sharing your practice? I am always amazed at the lack of information that is given to patients about some eye conditions such as AMD. Many relatives attending medical retina clinics with their relatives who have developed AMD are unaware that it can be hereditary, that it is related to age, smoking, poor diet and possibly eye colour. Yet we make provision for the relatives of patients who are being treated for glaucoma, by offering free optometry check-ups after the age of 40.

Despite anti-smoking debates and environmental limitations recently imposed on smoking, even the 'five a day' campaign for example, I would vehemently argue that we as ophthalmic nurses have a huge role to play in informing and advising our patients about eye health. Whilst in my particular speciality (the Medical Retina Service) there is much research under way pertaining to treatment and stabilisation of vision, I still advise patients about vitamin supplements if their diet is questionable, giving up smoking and generally taking better care of themselves. I am not yet convinced that we are doing enough to educate patients or the general public about eye conditions before they occur. It is now not beyond the realms of possibility to identify groups of people who may be predisposed to certain eye conditions. I truly believe that we should be advising them how to monitor their eye health and to seek help sooner should they notice that their eye health is suffering. I would like to see stronger links between ophthalmic nurses, Sensory Impairment Teams (Social Services) General Practitioners and Walk-in Centres, where we provide these health care providers with information leaflets or arrange to do teaching sessions to educate and advise them about susceptible patients. If you are already doing this, share your practice with other Trusts, we could learn from each other in this respect and work towards standardising the service we offer. Sometimes this will be a painstaking process, but the rewards in terms of patient care will prevail. The flip side to this argument though is, once again, you can only *give* information which allows the patient to choose whether to make appropriate changes to meet their eye health needs.

Let's use Age Related Macular Degeneration (AMD) as an example of how we could move forward in this area. AMD is one of the leading causes of central vision loss in the over 70 year old population. Since this is an age-related condition that has been linked to smoking, poor diet, heredity and arguably more prevalent in blue eyed individuals then one would assume that we have developed a very slick information and referral process. I can say that this is not the case. As a specialist field are we responsible? Perhaps we need to take the initiative and do something about this? In my particular area (Retinal Services), resources have not yet allowed our AMD Service to be launched fully. However all general practitioners and community optometrists have been informed about our services. This has been done by notifying them about referral criteria guidelines, which has meant that many patients are not waiting as long for an appointment, which is great. However, the focus has been cure driven rather than being predicated on prevention. Notably, patients suspected of having AMD are referred directly into our PDT service either to be filtered to identify whether they have the exudative (wet) form of the condition or they go through to our rapid access investigation clinic to assess the lesion type and treatability. This is one way that we can educate patients and their significant others about the condition process, especially raising awareness with those patients who have only one affected eye. However, one could argue that we are still not getting to the root of the problem by advising patients to check their vision regularly before they even develop symptoms or to raise awareness when they are entering the at risk phase, which in this case is largely old age. Could it be that we are relying on our optometry colleagues in the community to convey this information?

You may be thinking if every patient with suspected AMD was referred into the department, the floodgates would burst let alone open! We need to filter out those patients who do need to be seen by a consultant and those who could be better managed in the community. There is clearly a way of monitoring such patients in the community. Just as there are stable glaucoma clinics, why shouldn't there be stable medical retina clinics, especially for conditions, which may recur? This idea has been bandied about for some time now, but it would be great to see how it would work.

I almost cried "Hallelujah" when AREDS (Age Related Eye Disease Study) was first discussed. 'Over reactive' you might think but for me this was a way that I could involve patients in their own care, not just by taking care of themselves generally but by advising them about studies and the potential value of their findings. This was a great piece of research in terms of my practice. It was by no means conclusive, but it was useful and encouraging. As I am sure that you are aware, The National Eye Institute stated that the use of certain vitamins and anti-oxidants in particular may play a role in helping people at high risk of developing advanced AMD (dry type) to maintain their vision. This along with other antioxidant trials, with lutein and bilberry extract, finally meant that patients could participate in their own care, where dry AMD was concerned. I believe that it provided a psychological boost for both my patients and me. There are numerous leaflets that advise patients of this and how to tell if they have wet or dry AMD, so advising patients in this respect and keeping abreast of developments in your respective specialities is clearly beneficial to all parties both psychologically and physically. Many patients want to contribute to their own care, and for them this meant taking some control and responsibility over their care. I would argue that whenever you encounter a patient who has deteriorating central vision they are generally pretty open to any form of treatment, even intravitreal injections, which may have the effect of stabilising and in some cases now improving their vision. Given that they have already experienced some vision loss, they will be keen to maintain or reduce further loss of vision.

On the other side of this discussion, let's think about the young patient with diabetes who is not controlling their condition well and is

Challenging Assumptions in Ophthalmic Nursing:
a patient centred approach

constantly running at a blood sugar level of between 15 and 25 mmols. At present they are not really experiencing any major problems with their vision and your pearls of wisdom have fallen foul to indifference, thereby having little impact on their attitude towards their systemic health. This notion of denial is common and I am sure that you will have experienced it on many occasions. How can we as nurses make any kind of impact in such situations? If you think that I have the solution to this problem then think again! It is an age old problem and each situation will merit a different approach for each presenting patient.

You can only advise and inform the patient in a non-threatening, non-patronising way. Patients, in my experience, particularly those with diabetes, which is poorly controlled, are more than likely going to demonstrate resistance and display defensiveness. How can you tell a patient with fully functional vision that they will ultimately lose vision if they don't take better care of their sugar levels? How many times have you heard patients say, but it's normal for me to run at, for example, 15mmols? Do you have a good reply to that? As nurses we know that our bodies function in basically the same way and when something goes wrong we need to return the body to status quo, which is not 15mmols in terms of blood sugar. What you need to have is evidence. Perhaps you have already thought of recruiting patients who have lost sight due to diabetes to come and talk to these types of patients. Remember though that these patients, especially when they were newly diagnosed, spent more time than they would have liked in hospital and around health care professionals. They are often plain fed up of being in such environments and would like to escape as quickly as possible. How can you make their ophthalmic experience better and less intrusive? It's not an easy thing to think about, but given the huge proportion of people with diabetes that go through eye departments perhaps it's something that seriously needs addressing. Do you hand out leaflets about associated eye conditions? Do you have a specialist nurse who you can use as a resource? Do you identify patients of concern to the specialist nurse or even to the ophthalmologist?

Smoking was mentioned earlier. How do you tell patients about the links between smoking and vision loss? What do you say to someone who has been smoking for most of their lives that they need to stop to prolong their sight? Patients in this situation may respond by saying that they don't feel that it's worth giving up smoking now, despite the supporting evidence that you present them with. I don't know about you, but on many occasions I have been party to discussions about nurses who are smokers. They go for their break, returning smelling of cigarettes and then begin to advise their patients about the benefits of stopping smoking. It's a tough one, being a non-smoker my slant on this topic is this, you have to do what is in the best interests of your patient. If you are advising them based on their eye condition and the implications of continuing smoking they will draw their own conclusions and at least be able to make an informed choice.

We are probably all familiar with the famous idiom, "You can lead a horse to water but you can't make it drink". Without meaning to demean our patients, because let's face it our patients could be anyone, this adage exemplifies my encounters with some of them. There is clearly a disparity between how you are able to 'educate' patients with different eye conditions and how well that advice will be received. I am about to make a huge generalisation but indulge me. Let's think about a patient with AMD and a patient who presents with non-proliferative diabetic retinopathy, I would contend that the patient who has AMD would demonstrate a greater degree of receptiveness than the other patient. I believe that this is probably because the former has already experienced some vision loss and the latter has not. However, there are still patients with diabetes who have lost a lot of functional vision and are still not willing to adhere to dietary changes required for their eye health, and unfortunately there are those who have adhered but are still suffering. There are also patients whose natural make up may predispose them, such as heredity, smoking, eye colour, environmental climate, to developing AMD or other eye conditions.

So why bother telling your patients about eye health care? As we identified earlier, the emphasis is on prevention rather than cure.

However there are new treatments being licensed and made available for ophthalmic use, which may treat or stabilise vision loss. One might contend that the availability of such drugs may have the opposite affect on patients and the way that they think about their health. What do I mean by this statement? I would argue that there already exists a group of patients who may choose to live now and worry about things breaking down and wearing out when they actually happen, in the hope that at that point, there may be a remedy of sorts available. We are living longer and arguably remaining independent for longer because of modern medicine. Many of my more mature patients are aware of newer treatment modalities. Most of these people have often lead full independent lives and are desperate to continue to do so. Rather than losing their vision some would subject themselves to intravitreal injections every 4 to 6 weeks without any real knowledge of an end to the treatment regime. They would be prepared to subject themselves to a combination of therapies if it meant just a few more years of better vision, exposing themselves to risks of endophthalmitis, haemorrhage, raised intra ocular pressure and so on. Whilst these are not common complications, they are potential complications none the less. One could argue that health advice for these patients is pointless, but vision stabilisation may still require some life style changes, improved diet, vitamin supplements, better blood pressure and cholesterol control. I would argue that health advice is an integral part of our roles and should be given readily to all patients. I am not advocating becoming broken records. We should be sensitive to patient needs and their emotional status, identifying high risk patients and making health advice accessible through leaflets and discussing health issues and outcomes.

References

- The National Eye Institute
  http://www.nei.nih.gov/amd/summary.asp)
- AREDs Study information - http://www.smcok.com/media/newspaper/november/ared1103.htm
- AREDs Study - http://www.countrydoctor.co.uk
- AREDs Study - http://www.sciencebasedhealth.com
- AREDs Study - http://www.ophthalmologymanagement.com/article.aspx?
- About stopping smoking - www.smcok.com/media/newspaper/november/ared1103.htm
- Government recommendations for healthy living - http://www.dh.gov.uk
- Smoking and Vision Loss – RNIB site http://www.rnib.org.uk

Challenging Assumptions in Ophthalmic Nursing:
a patient centred approach

Patient Compliance

Have you ever raised your eyes skyward when your patient does not:
- Attend at the appointed time or even attend at all?
- Use their eye drops or other ophthalmic prescriptions (whether topical or systemic) either in the way instructed or at all?
- Comply with some or all of your express instructions about their follow-up or after care?
- Tell you when they don't really understand what it is you've told them?

If you said yes to one or more of those did you ever ask the question as to why that might be?

Patient compliance or non-compliance as the case may be, is a very complex issue and not as straightforward as it might seem on the surface. Let's try a bit of empathy here and see what we come up with. Empathy is defined by the American Indians as standing in another Indian's moccasins and viewing the world through his eyes. In other words trying to think about how you might feel if you were in the same or similar situations. It is a concept which you may have come across during your nurse training but in my experience there is not a huge amount done on empathy. It's almost like we should just have it without really exploring the emotional underpinnings of it. There is a belief in some quarters that empathy is something which develops over time and with experience; it's not something that can easily be taught. However, it shouldn't be confused with sympathy which is a completely different concept and if you're not sure of the difference then you need to look it up in a dictionary! All too often we're caught up with the here and now and trying to think two jobs ahead to pay much heed to the patient we are sending home with a shed load of instructions and medications following some ophthalmic procedure.

Consider this scenario. You are 75 years old and being discharged from hospital following surgery. It can be any surgery you like maybe a cataract or some such thing, that bit isn't important in the greater scheme of things. So your husband/wife who is of a similar age has come to collect you. Staff Nurse says she will give you your OPD appointment and go through your TTO's. Remember you are 75 and don't know your OPD from your TTO, jargon which we all too readily and easily bandy about with gay abandon forgetting that most of our patients don't have a clue what we are on about. You are unaware (and let's face it you probably don't care that much as you are going home) that the ward is two staff members short, for reasons which we will not go into here, but you can make anything up which bears a resemblance to real life.

You think the Staff Nurse seems a bit harassed but again this is not your problem and she is always a bit abrupt anyway so you tend to

Challenging Assumptions in Ophthalmic Nursing:
a patient centred approach

ignore it (does this ring any bells with any of you? Could you be that grumpy Staff Nurse?). You and your husband/wife have gathered up all of your worldly goods and are waiting, perhaps a little impatiently, for your final instructions. You keep checking that your husband/wife has got all of your personal belongings because you know what men/women are like. Your eye is still a bit uncomfortable and you're not really sure whether this is normal or for how long this is likely to persist. You are a little bit nervous about going home as you don't know if your husband/wife is going to be able to look after you, or more to the point your eye, properly and what you will do if anything goes wrong. These and other myriad thoughts are buzzing round your head. This in itself worries you because you might forget to ask all of the questions which are jostling for position. The Staff Nurse hurries over to you clutching a paper bag and a collection of what look like leaflets. She tips the contents of the bag onto your now vacated, but probably still warm, bed. There seems to be a multitude of pots and bottles spilling out over the counterpane. Now remember you've had ophthalmic surgery so your eyesight may not be all that it could be. Maybe you've even got a pad or some form of protection on one of your eyes. You are 75 years old. Are your hands as deft and nimble as once they were? Staff Nurse picks up each bottle each which seems impossibly tiny and points to the frankly unreadable dispensing instructions. Unreadable because the print is small so that it can be fitted on the bottle. That doesn't matter though because after all it is written down! She tells you about putting drops in your eyes but doesn't have time to show you or advise you on how this might be done. Ah, but there is a bit of paper with it all on. Shame you can't actually read it. Never mind your husband/wife will sort it out when you get home. As for manipulating the bottle and actually getting a drop out of it well not to worry we'll get the District Nurse to come in and assess you. If she's not in too much of a hurry she might even do it for you. The trouble is she can't actually get to you until next Thursday so you'll have to manage until then. Then Staff Nurse goes through the other medication. This one is for this and this one is for that and you must take this one with food and this one after food. OK? Your OPD appointment will be next Friday morning at 10.30 but can you come at 9.00 so that we can put your drops in. Well it's a bit difficult because your husband/wife can't bring you at that time on that day (again this could be for any conceivable reason so you can fit the one that best suits the purpose). Staff Nurse just holds off a 'tut' but hasn't got time to change it so you'll have to ring this number when you get home and sort it out. Any questions? No? You haven't even had time to open your mouth let alone say anything and you're not sure you understood anything because while you were listening you were also trying to remember all the questions you were going to ask. Your husband/wife looks just as bemused by the rapidity of the information. Bye bye then. You clutch the TTO bag with all of your medications and the various leaflets and appointment cards. Your hair is ruffled in the jet stream of the recently departed Staff Nurse. Not quite thrown out onto the street but damn near it!

This of course is an exaggeration, isn't it? Or is it? Can you honestly say that you have always diligently and meticulously gone through the small print with the departing patient? Have you used your communication skills to their maximum to determine whether the patient has actually understood the information overload? Have you fully determined whether they can indeed hold the drops bottle, take the top off the bottle, squeeze the drops bottle or even get the drop in a close approximation to the eye? If you have found that the patient can't do these things have you asked whether the relative/care can do these things? What if they can't only because they have never done it before? Do we provide instillation of drops classes to all of the relevant people whether they are the patient or a significant other?

In my experience, and I know this is a broad generalisation, but, patients do not set out with the objective of being non-compliant. Yes there are a few awkward customers who can be objectionable and will do the complete opposite of everything you try to tell them. But on the whole these are few and far between. Most people coming through our service do so because they want us to make them better and to achieve that end they do what we tell them to. Unless of course there is a very sound reason why they cannot!

Try this simple exercise in empathy. Put on a pair of fairly tight fitting rubber gloves. Next fill them with peas or sand or a similar product. In fact you could do the experiment using a variety of substances just to see what the different effects are. You need to make sure your hands are completely encased in your chosen material. Ensure the gloves are secured at the wrist so that the product does not leak out and to ensure you get maximum effect. Now pick up a drops bottle and open it. For the next bit you'll have to put the lid back on so that you don't actually inadvertently self-medicate. Hold your eye open and move the drops bottle into position as though you were going to insert a drop in your eye. Now squeeze the bottle to simulate the instillation. Next on a piece of paper rank on any scale you wish to choose, but either a numerical or Likert type scale will do, how easy-difficult each part of this process was to carry out. What you have just done is simulated someone with hand problems, something as simple as stiffness due to age related changes. You could also vary this experiment by wearing a soft collar and trying to tilt your head back sufficiently to get a drop in it. The permutations are endless and you can have 'fun' in coming up with ideas of your own.

Are you getting an insight (no pun intended) into why your patient may have difficulty with self-medication? Well, why don't they say anything?

Do you like to admit that you can't do the things you used to be able

Challenging Assumptions in Ophthalmic Nursing:
a patient centred approach

to because you're getting on a bit? I certainly don't and I don't know too may people who do. Chuck in a bit of ageism, the elderly feeling that they lack worth, and they definitely will not admit to any failings in their abilities. You can apply this to any group of people, the disabled, ethnic minority groups, learning disabled people and so on. The scenario we went through earlier is only the tip of the iceberg and could be replicated across any of your patient groups. Fear plays a big part in all of this too. Fear of doing it wrong; fear of making something worse; fear of losing their sight; fear of appearing stupid; oh I could go on for hours. Remember that when we are afraid we don't always focus on the task in hand. We might focus on one particular aspect of our care to the detriment of other aspects. This is natural. Each different fear that the patient has, and they may have many fears such as those mentioned above, will compound the other fears. There will be potential for the fears to saturate the patient's mind so that nothing else goes in. Cast your mind back to a time when you were afraid or anxious. Perhaps it was a job interview, your final exams, a nightmare, that doesn't matter. It is the fear which fills your mind, not the thing you're supposed to be thinking of. Did you go blank at the exam question or when the interviewer asked you a question? Do you remember the feelings of panic because you couldn't think of the appropriate answer? Do you get the picture?

What we really need to get to grips with is that prevention is usually much better than cure. If we were to start as we mean to go on right at the very beginning, maybe even before the actual admission then we could potentially overcome some of these so-called non-compliance issues.

Some ophthalmic units now have pre-admission assessment units. Some just have them for surgical admissions, some for medical, some for both and some for none of the above. Pre-assessment units generally deal with issues of medical fitness for treatment. Some of the more forward thinking or pro-active units start to plan the discharge at this stage before the patient has even hit the unit as an in-patient. Here the patient's general abilities to man handle items of equipment such as drops bottles or medicine bottles can be assessed and plans made to provide aids or education and training to the patient and/or the carer. At this point questions could be asked as to the ability of the patient to return to hospital following treatment. Perhaps they did not attend (DNA) because they experienced transportation difficulties. Perhaps they were only able to get a lift when their neighbour/child/partner comes home from work. Many people working for fairly generous companies such as the NHS don't realise that people working for less generous companies do not get paid if they don't go to work. So maybe the person giving the lift cannot take time off work because they will face a financial penalty. We professionals make a lot of assumptions about the people we care for based upon our own experiences and maybe those experiences do not reflect the experiences of our patients. If your workplace does not have the luxury of a pre-admission unit, then planning has to start when the patient is admitted. This may be no mean feat if the patient is booked for a day procedure but it is essential if the patient is to be as compliant as they possibly can.

Take a look at your nursing documentation. Does it ask searching questions of the patient? Does it provide enough information for you to be able to ensure the best follow-up care, tailor-made for the patient and his/her carer is provided? Do you check whether they/the carer can hold and operate a bottle of drops? Are you 100% sure that the patient/carer understands the supreme importance of using the drops/medication and for attending for follow up. Or is that just another assumption that we make? As I said before, we make a number of assumptions about our patients and they are not always the correct ones! People do not always understand the importance of complying with instructions, particularly if we haven't told them.

So next time you have a patient who hasn't fully complied with instructions they have been given about their care or treatment, don't just 'tut' and pull a disapproving face, look deeper. Maybe there are sound reasons why they are not compliant and you have the responsibility to unravel those reasons however trivial they may appear on the surface.

Challenging Assumptions in Ophthalmic Nursing:
a patient centred approach

Evidence Based Practice and Research

On a scale of one to ten, without crossing your fingers behind your back, how evidence based is your ophthalmic (and other) nursing practice? If you said ten then well done, but I would suspect that for most of us it is somewhat less than this. That's not to say that you don't use evidence-based practice (EBP) because it is likely that you do to a greater or lesser degree. A lot of the practices and procedures that you carry out in your every day nursing care are probably evidence based at some level. However some are also ritualistic practices that we have done because we were told that was how it was done or because it has always been done that way. The medical profession claim to be very much evidence based in their practice, but I would suggest - perhaps tongue in cheek, perhaps not - that this is not always the case. An unpublished research thesis carried out by Liggins (1999) showed that only a very small percentage of the medical profession actually used EBP even though the majority claimed that their practice was evidence based. If you have ever worked with surgeons you may have seen this. Surgeon A uses this set of instruments, prosthesis, sutures and dressings. Surgeon B does the same operation but uses a completely different set and in a different order. You may have worked in areas where there are similar experiences with nurses. Nurse A does a procedure this way and Nurse B does the same procedure this way. Who is doing it the right way? Is there a right way? Which way is EBP? The problem is that both ways may be evidence based. The question may be how long ago was it evidenced and whether it has been superseded by more up to date evidence.

To explore this issue further we really need to address the reasons for EBP. I can see you raising an eyebrow at this. Surely the reason for EBP is to ensure that the patient gets the best possible care. That's probably, and should be; quite true but there also may be something to do with professional standing here. EBP has a certain ring and makes it sound – to the uninitiated - as though everything they do is firmly steeped in the best available evidence. It is easy to blind (no pun intended) patients with acronyms, abbreviations and a lofty approach. Using these techniques can make anything sound plausible and can be a useful way of implying professionalism. However, I come back to the very first question. How much of your professional practice is evidence based? Before the days of aural temperature monitoring how may of us left the mercury thermometer in the mouth for the required 7 minutes? Yes 7 minutes, that's what the research said was the time required for an accurate recording. Most nurses I know bunged the thermometer in the mouth, grabbed the wrist counted the pulse for at most half a minute (sometimes only 15 seconds if in a hurry) did the required mathematical calculation, removed the thermometer and recorded it on the TPR chart. No more than 2 minutes in total and hardly evidence based. None the less we probably were all guilty of it at some level. To be fair though, there were probably not that many of us who were aware of the fact that we needed to leave the thermometer in for that long. Can you imagine what it would have done to the TPR round?!

And there is the rub. EBP can seriously get in the way of current practices, procedures and dare I say it 'rituals'. It has the potential to seriously interfere with established routines that we have developed over years to get the work done in the amount of time we have available to us. It can also take us out of our comfort zones. We get the hang of a procedure and get to the point where we can do it without having to think too much about it. Suddenly some bright spark comes up with a new way of doing it and you have to learn it over again. Maybe because we don't like change we are slow to embrace this new way and when no one is looking we revert back to our preferred way. Unless we are being constantly supervised we can continue in 'the old way' unhindered and so our practice stays the same and doesn't move on at the same pace as the evidence. Another issue of course is which bit of evidence to embrace. A piece of research is published extolling the virtues (or otherwise) of certain practices. Being EBP minded we take it on the chin and put this into practice. We work hard to do this only to find a piece of research is published that runs counter to that which we have worked so hard to implement. What do we do? If we are truly into EBP then we should change. After all we should not expect practices to stay static. They should be dynamic, ever evolving to meet the rigours of academic and scientific requirements. Ultimately we should be acting in the best interests of our clients, the patients. We wouldn't dream nowadays of keeping patients who have had a cataract removal more that a day. Not so long ago we would keep them in hospital for up to (and sometimes beyond) a week.

These changes though can be extremely difficult to manage, after all it's not like you've got anything else to do in your working life is it? This means that maybe we end up being selective about our EBP and the types of practices we base on the available evidence. Sometimes these changes seem to appear as if by magic. One day you go to work and do something one way but when you go back after days off perhaps or annual leave it has changed. But you never seem to find out who said it must change or why it must change. So how do you know it truly is EBP and not the whim of some passing doctor or dare I say it, Manager? Very often we never do get to find out but maybe that's because we never ask. When you go to work and you carry out your nursing practices and procedures do you ever in reality, stop and ask either yourself or your colleagues what the evidence is for it or where it is published? Maybe, if you are doing a course, you will pursue this and find out where it originated but in general I would suggest you will carry on regardless until someone tells you to do it another way! I make no apologies for making that challenge as in my experience this is exactly what happens. We are too busy or dare I say simply not interested in doing things differently even if it is better for patient outcomes. It is much easier to 'put up and shut up' as the saying goes. We all want an easy life but should that easy life be to the detriment of our patients?

If you feel this is unfair then consider this, can you hand on heart honestly say that you read all the available evidence for the care you carry out? There's plenty of it about so there is no excuse for not being able to access it. Also with the internet, evidence is widely available through the various databases that most university libraries host. The problem that some, and I use this term loosely, professionals have is with the academisation (yes I know it's not a real word but it says what I mean) of nursing. Practising nurses are very sceptical about the need for and the value of academia and research to underpin their practice. Some but by no means all take the stance that them that 'can', 'do' and them that 'can't', 'teach' (and them that can't teach, teach teachers to teach)! The feeling being that if you are any good at nursing then that's exactly what you do, you nurse. The people that can't hack it either go into management or even worse go into education. From my own experience, I was in a far better position to influence care when I was working in Nurse Education than when I was a practitioner, yet still I was treated with suspicion because, according to my practising colleagues, I didn't know what it was like in the real world. This just proves to me how little they know. This might be an unfair portrayal, even a stereotypical view but it is one that is based on personal experience.

In addition to this there are still nurses out in practice who view the new diplomate and graduate nurses with suspicion. I do have some sympathy with them as nursing did itself no favours with the early university curricula leaving practitioners not fit for purpose (UKCC 1999) But we have surely come on leaps and bounds in the 10 or so years we have been affiliated with the universities.

Universities have given weight to our professional status. We are up there among the other Allied Health Professionals (AHP's) now whereas before we perhaps trailed behind if not in our eyes certainly in their eyes. Universities are producing knowledgeable, questioning practitioners who want to underpin their practice with evidence. They also want to add to the evidence base themselves as more and more nurses become research-minded. Most, in fact I would go so far as to say all nursing courses, both pre and post registration and including the Ophthalmic Nursing Course are very much evidence based. They encourage enquiry and a questioning approach to nursing care, which is what we want to engender and which, is one of the reasons we wrote this book. I hope you will go into practice and ask why? Where is the evidence? Who says? And so on and so forth. Eventually others will join you as in my experience the old adage 'from tiny acorns mighty oaks grow' is actually very true.

So what can we do to ensure that evidence really does form the basis for our practice?

Challenging Assumptions in Ophthalmic Nursing:
a patient centred approach

Well, there are a number of ways some of which you may have already used or come across.

The Link Nurse is a great way of keeping the team abreast of developments in different areas and you may have these beings in your own area. The Link Nurse is someone who may have a special interest in a particular aspect of ophthalmic nursing but who works generically rather than as a specialist practitioner. He/she is the one who attends the appropriate meetings where current practice is discussed. The Link Nurse can then go back to their work area, disseminate this information, and help to implement any changes that are needed. The Link Nurse becomes the crusader for change and should be instrumental in ensuring that the ward/department adopts the most recent and up to date practice.

Ward meetings are also a useful way of looking at current practice and current research findings and discussing ways in which care can be enhanced. Team members can bring ideas for discussion maybe as an extension of the Link Nurse role. Obviously the Ward Manager has a huge part to play in this, as he/she will be the driving force for change. However if your manager is less than keen it doesn't have to stop you. Make EBP an agenda item and feed it into the meeting. Very soon your colleagues will become used to the idea and may start to look for EBP for themselves to bring to the meetings. Get one (or all) of the Ophthalmic Specialist Nurses on your side. Ask them to come to the ward/department meetings to give a brief talk.

You could also become the ophthalmic EBP Champion, flying the EBP flag wherever you go. The downside to this zealous approach though is that your colleagues may start to avoid you so you have to not over do it. You could however ask your library to stock the ophthalmic journals (if they don't already) and copy appropriate articles, making sure you don't break the copyright laws of course. Check out the stock of ophthalmic books that are held in the library. How up to date are they? Is there a supply of nursing literature rather than just medical literature? Is this book on the shelf? You can always lobby the librarian so that when they update their stock you can suggest that there are relevant ophthalmic texts available.

Student nurses are always a good way of keeping abreast of developments, so if they are in your area use them. If you don't have them why not see what needs to be done for your area to become a placement or part of a placement for them. Because they are exposed very early on in their training to basing their assignments on available evidence, they are in a prime position to help you out. If you don't like the thought of asking a student nurse because you don't like to show your ignorance, then that in itself is a good incentive to go and find out for yourself. But that apart, they are a very useful resource and they will question, as part of their own learning, the practices and procedures that you are undertaking.

Challenging Assumptions in Ophthalmic Nursing:
a patient centred approach

Don't forget the post-registration nurses doing the ophthalmic course. Talk to them, listen to what they are doing on their course ask them to give a talk at a team meeting. They should be developing an enquiring approach to their ophthalmic practice and should be able to help you develop yours. Ask to see their assignments. Is there anything you can help them with or get involved with as part of their course?

A forward thinking member of the medical profession can also help your cause especially if they can see a benefit to them and their patients as a result. You could also potentially get involved with research that many of them undertake then you really will be at the forefront and you never know you may get your name in lights as well.

You could also get yourself on an ophthalmic nursing course. There are a fair few around and about but you may have to travel. There are also universities that offer an open learning approach, which may be useful if you can't travel or the department is too thin on the ground to release you for a study day. 'Google' ophthalmic courses and you may well be amazed by the results you get.

However, a note of caution needs to be injected here. Remember earlier I said we were slow to change as a breed and indeed many of us don't like to be pushed outside our comfort zones. So we need to be careful if we want our colleagues to change to new ways of practising. In my experience you have to adopt a 'softly, softly catchy monkey' approach to change. If we bulldoze people into doing things they dig their heels in and won't budge. It is almost a school kid mentality and it is something we are all probably guilty of doing. If someone tells us we have to do something we don't like it. If someone asks us whether we think it might be a good idea and should we just trial it and see we are usually much more receptive. What we want to do is to create a line of least resistance. Our work lives are hard enough as it is so we don't want to make them any more difficult. There will however always be at least one fly in the ointment and you will never win them round. But if everyone else is on board you can deal with the one that isn't.

It is not always going to be an easy ride in your quest for EBP but it is worthwhile and it is for the patient you are doing this. Don't give up; it will be worth it in the end. Just you wait and see.

References

- Liggins K (1999) Use of Evidence Based Practice. Unpublished Thesis
- UKCC (1999) Fitness for Practice. UKCC. London

Challenging Assumptions in Ophthalmic Nursing:
a patient centred approach

The Specialist Nurse –
Pushing the Boundaries of Ophthalmic Nursing Practice?

The expanded role has been widely discussed. Nursing practice has evolved or has it? Have we done full circle? Are we merely meeting the junior doctor deficit imposed by working time regulations as a result of Europeanization or are we truly becoming new entities, bridging the gap between Doctor and Nurse? What are the legal implications of this role? Where exactly do these nurses fit in to this system? What type of autonomy does this role actually have? Has this role paved the way for the development of the roles of general nurses within the speciality? Do we feel constrained by Trust requirements to ensure theory and practice match? There are endless Patient Group Directions (PGD's) and Protocols which, one may argue, limit practice but we are cognisant that they are also there to standardise care and to ensure patient safety. I want to challenge the perception of the expanded role and discuss its implications in terms of practice and development.

We have been and still are called many things in our evolutionary state, Consultant Nurses, Nurse Practitioners, Advanced Nurse Practitioners, Specialist Nurses, and Clinical Nurse Specialists to name probably the more commonly associated titles with expanded practice. This is clearly not exclusive to ophthalmic nurses but it's a discussion worth having. I want to start this discussion by sharing my experiences as a specialist nurse (Clinical Nurse Specialist - CNS) because I want you to know that you are not alone in some feelings that you may have experienced or are in the process of experiencing. Being innovators in practice tends to set specialist nurses apart from other nurses whether we like it or not and I want to acknowledge that expanded practice comes with both positive and negative connotations.

The job description for my role as Medical Retina Nurse Specialist was seen as developmental from F to H (old grading). It described acquiring skills such as fundoscopy (examining the back of the eye using a special lens), digital image assessment, running the angiography service, running the photo dynamic therapy service (this therapy is used for some patients who have a certain type of wet macular degeneration), tonometry (checking the internal pressure of they eye using a special apparatus), counselling, teaching and research to name but a few. I am now in the next phase of my development and much of what the role requires now does not include skills previously required. I made the role into what I felt the patients needed. The guidelines that were originally in place were exactly that, just guidelines. Fortunately I had an amazing manager who helped me to shape my role, but one of the things I found most difficult was the isolation and the autonomy of the role. Things which ironically I now treasure!

Challenging Assumptions in Ophthalmic Nursing:
a patient centred approach

I had to sit down and work out what my service needed. This turned out to be something entirely different to the vitreo-retinal service, the service that I was supposed to back-fill. The first six months of the post were the most difficult. I continued to work in the department where I used to be a staff nurse, wearing my sister's uniform, and staff did not understand what I was supposed to be doing. I was no longer line-managed by the manager of that unit. I had a counterpart in the surgical side of the retinal service and I had to back-fill her as required. It was probably one of the most confusing times I have ever had. If I had to describe my specialist role at that time to one of Kim's Martians I would talk about things like practice and service development, expert practice function, innovative practice and so on. Pinning my role down on some days has felt rather ethereal. There is much debate about whether there should be a codification of specialist titles so that it is instantly understood what each title means. There is a clear need to do this as I have recently discovered when looking at the Knowledge and Skills Framework (KSF) requirements and the job description for my role, but based on what my counterparts are doing elsewhere I am also not entirely convinced that this would produce the desired outcome. However, being able to identify with a role should enable staff to know what each person's remit is within a department. Having the confidence to shape a role and be dynamic within that role will help with this process and is truly in keeping with the current climate of ethical, legal and efficacious practice.

Often nurses are the most experienced resources (apart from consultants and resident senior registrars of course) new doctors have when they come into an ophthalmology unit. Doctors regularly rotate through our department and often rely on nurses for advice about treatment and investigation pathways. I have known photographers, optometrists and nurses wave an angiography form under the nose of a doctor saying "this patient needs an angiogram". All parties know that this procedure needs to be prescribed by a senior doctor or consultant (in our department) because fluorescein sodium is still an unlicensed drug, yet it often happens that a form is signed based on the doctor trusting the knowledge and expertise of the nurse or photographer. At this point I will add that I am aware that in a couple of Trusts this practice has been formally authorised. A Patient Group Direction has been passed to allow certain nurses to prescribe fluorescein sodium for angiography purposes. One may justify this practice as sight loss prevention. Acting quickly ultimately reduces the waiting time for a diagnostic procedure where vision is deteriorating thereby increasing the potential for early treatment. The patient may thank us but are we pushing boundaries? It seems that Trusts will make some allowances where nurses have to work beyond their remit. If this is happening in some Trusts why is it not happening in all Trusts? Is the absence of experienced medical staff the reason? Is it because nurses in some areas are more experienced and competent to make such decisions? Are we moving faster than the law? If so, what if something goes wrong? Medical

Challenging Assumptions in Ophthalmic Nursing:
a patient centred approach

litigation is growing and we all have to justify our practice, produce clear evidence and records of care and ensure that we all stay within our respective organisational "sphere of scope and practice" (NMC). Whilst we are covered by vicarious liability under law, as nursing roles expand perhaps we should be thinking about getting more cover in terms of liability. We are seeing a steady increase in the number of litigations relating to medical negligence or malpractice these days, which demonstrate a corresponding increase in the amount of monetary compensation awarded.

Many experienced specialist nurses could intercede without asking a doctor and may arguably have the knowledge and skills to assess the patient's images, from digital photography and optical coherence tomography (OCT) to determine whether a patient may benefit from angiography. In some Trusts, protocols have been developed to allow this practice. In the meantime in Trusts where this is not common practice, should we be content that we can influence doctors to prescribe it? Not that we are able to do this. I am merely playing devils' advocate. There are other instances where tried and tested drugs are being used to treat patients, and procedures are being performed on patients where specialist nurses who have been assessed as competent may be able to intervene to prepare the patient for one visit to the consultant as opposed to two or three. Any specialist nurse will tell you that their practice, particularly in terms of procedures, is determined by a strict set of rules codified in a protocol or patient group direction. I wince when someone mentions writing protocols. The very thought of spending time writing one makes me feel decidedly unwell, but these are the bread and butter of our service. They standardise practice and ultimately improve safety and efficacy of patient care. This apparent ambivalence between pushing boundaries and ostensible limitations may make you think, what is she on about? Indulge me here. The only way that specialist nurses get to do procedures is to be trained by doctors or optometrists. I would argue that once trained, specialist nurses are in a fantastic position to not only codify their training but then to review and influence practice, making it more transparent and being better able to advise and support patients. This may seem obvious to most of you but some may only see the limitations of protocols. I personally feel that protocols and patient group directions are positive tools to protect the patient, standardise care and to allow the nurse to give patients the care that they deserve. Leaving aside individual personalities (of specialist nurses), in my view, patients entering a nurse-led service know what they are going to encounter and they receive the same quality of care each time. Nurse-led services have this habit of demystifying procedures and treatments that the very names of have instilled fear into the hearts of patients alike. I realise that this is a slight exaggeration but in some respects there is a truth here. I would proffer that the emphasis in these services is information giving which has the effect of allaying patient anxieties for the most part, which in turn has the effect of

Challenging Assumptions in Ophthalmic Nursing:
a patient centred approach

streamlining services.

The blurring of boundaries between medical and nursing practice has been a dominant theme in many discussions about the expanded role of the nurse. I suppose one could contend that sometimes it's difficult to know where the experienced nurse starts and the junior doctor finishes. There is this notion that the nursing role is becoming over medicalised. What do you think? Do you feel that we are becoming more technical? Personally, I think that in some respects we are, but I would argue that in developing certain skills we can better serve our patients and reduce waiting times. I would also assert that at the heart of everything a specialist nurse does are the best interests of the patient. We must remember that whilst we are focused on patients as individuals, each nurse has his or her individual qualities, interests and idiosyncrasies. To say one is a specialist nurse conjures up all kinds of images and for my part having undergone the process of becoming one, I would have to say that my perception of my role and how I have been integrated into the scheme of things has altered considerably. Each nurse whilst following standards set by each Trust will ultimately develop systems to achieve best practice for their respective group of patients.

How then can we validate this role? How can we truly capture what it means in each situation? For example in our department alone we have four specialist nurses, two for retinal services, one for ocular plastics and one for glaucoma. We each have very different roles and different styles of performing those roles. Whilst one may argue that we do not perform as quickly as the junior doctors, in terms of how many patients we see, patient satisfaction in each respective service is high. Ongoing audit is crucial to this role, in terms of determining patient satisfaction, standard setting, cost effectiveness and justification. At one point we were asked to submit diaries containing our duties and the number of patients we see in each session we hold. There were always going to be hoops to jump through to prove that we were worth our weight in gold!

Agenda for Change has gone some way in formulating a set of core criteria for specialist nurses across the board. The RCN have been working hard on developing ophthalmic nurse competencies, perhaps we should go one step further? Perhaps we should network with other nurses in our particular sub-speciality, such as medical retina nurses, vitreo-retinal nurses and so on and agree core objectives for that role with a career pathway structure? Is it possible to do this, after all each Trust has different staffing levels and requirements? I think that it is certainly worth exploring the possibility of developing core responsibilities nationwide. However, I would argue that the respective roles of specialist nurses are still in a state of flux. We still seem to be undergoing a period of adaptation and evolution, based on our patient and practice needs.

Challenging Assumptions in Ophthalmic Nursing:
a patient centred approach

Why bother having specialist nurses? Surely services would run perfectly fine without them? I have heard the comment that clinics run just as well without us. Despite this I recall patients saying, we missed you on our last visit, no one seemed to know who we could talk to about our needs. In my experience most consultants prefer to have their specialist nurse with them in clinic for all of the reasons previously mentioned. If you are a general ophthalmic nurse reading this, think for a moment about what you believe the role of the specialist nurse is in your clinics. Do you use him or her as a resource for information? Do you refer to him or her when you are uncertain about a patient? Have you ever asked them what they do and how you should relate to them? Have they ever explained what their role is to you and if so, did you understand? I would challenge you as general nurses to get to know your specialist nurse and use them as a resource. It is their knowledge, skills and experience that places them at the heart of their respective sub-speciality. I would challenge specialist nurses to make their respective roles more transparent to allow others to identify with the service that they are providing to allay any potential ambivalence about their function.

So let's imagine the clinic without a specialist nurse. We are in a busy outpatients department running a medical retina clinic, and you are one of the general ophthalmic nurses on duty. You test the visual acuity of your patient and instil the dilation drops as per your patient group direction, you notice that the patient's vision has dropped significantly since their last visit and you make a point of highlighting it in the notes. You have another 30 patients coming in so you put the notes where the doctors will collect them and move on to the next patient. If you had a specialist nurse you could have asked them to see that patient. They would have been able to discuss referring them to the Social Services Sensory Impairment Team, or Low Vision Services, or they could have ascertained a more detailed history, ordered OCT or digital photography and discussed the case with the consultant or senior doctor. You may have had the time to do some of this but at least in this situation you would be able to rest assured that your patient would be well supported. It's such a simple thing to do but it requires time, which you don't always have. Utilising your specialist nurse in this way would provide the patient with a better standard of care and support than they would have probably ordinarily received. I apologise for teaching 'grandma how to suck eggs' but I can only base this on my own experience and discussions with other specialist nurses. Your department may be blessed with an Eye Clinic Liaison Officer, often funded by the RNIB (Royal National Institute for the Blind), who could deal with the Low Vision and Social Services part of this problem, but the skills and knowledge that your specialist nurse has would bridge the gap.

One of the frustrating parts about being a specialist nurse at present is that you may be asked to 'fill in' when there are staff shortages. This has been widely discussed in papers. I am not employed by the

Challenging Assumptions in Ophthalmic Nursing:
a patient centred approach

department I work in, but in the past I have been asked to help out when staff are off sick from my clinics. This puts my other duties on hold and makes me a rather expensive substitute for a staff nurse. This in effect can lead to confusion about the role. If you are seen to step in to help, your role is rarely taken seriously, but what actually happens is that you are still partly functioning in the same way. If I see a patient who meets the above criteria when I am taking their visual acuities, I will step in and do my role too regardless, otherwise that patient loses out. It's difficult but it has to be done. I sometimes feel that the way that patients are booked into the Out Patient Department causes what I usually describe as the 'conveyor belt effect'. Let me elucidate, patients are called in one by one, to have their visual acuities checked and drops instilled and then once seated in the waiting area, nothing happens to them (from the nurses point of view). I hasten to add that this is not directly a nursing issue. Often patient numbers do not allow more than a quick visual assessment and instillation of dilation drops. There is naturally, therefore, more emphasis on getting them through the system as opposed to identifying any special needs they may have. Obviously not every patient is going to need such attention and in my experience they and their relatives, want to spend as little time in the department as possible. Unfortunately I often get to see the patient after the consultant has seen them rather than utilising the time before, which may have the effect of reducing waiting times and allaying anxieties. However, I accept that there are times when the patient will need to see me after the consultation to discuss what the doctor has told them. Just take a moment to consider if it was you or your relative waiting, wouldn't you rather talk to the specialist nurse and ask questions prior to seeing the consultant? The nurse could write key points, go through the referral options, and discuss your eye condition and so on, reducing the consultation process. This also means that the patient has a link with the specialist nurse should they need further advice and support.

I have noted that specialist nurses actually perform some of the roles traditionally performed by doctors. Judging from the discussion in the previous paragraph I would argue that that is not such a bad thing, but then I am biased. Rather than dismissing this claim, let us examine it. Part of the discussion hinges on the fact that junior doctors being supplanted by specialist nurses could lead to junior doctors becoming de-skilled in some areas. I would contend that this is fallacious. There is generally only one specialist nurse in each area, unless you have a huge unit. In my humble opinion, it is a rather parochial view to hold, given the nature of the specialist nurse role. It is granted that specialist nurses perform some of the junior doctor's role but the specialist nurse is not task orientated, they are highly qualified in terms of communication, technical skills and the knowledge required to perform their respective role. Here follows another soap-box moment! We are not mini-doctors! We provide a service link between doctor and patient bridging the information gap and drawing in external links from the community to ensure that our patients with visual impairments are being well cared for and supported. Collaborative working is truly reflected in this role. I quite like the notion that specialist nurses are considered as "hybrids"... "striving to distil the best of nursing with the best of medicine" (Tye & Ross, 2000).

One could postulate that the introduction of the specialist nurse has been a great boost for the nursing profession. I just want to clarify that I am using the word 'introduction' loosely; specialist nurses have clearly been around in various guises for many years. I would argue that it has made nursing more appealing, from a career progression standpoint, giving nurses something to aspire to. The specialist role requires creativity, dynamism, a commitment to quality care and service provision. It encourages nurses to increase their knowledge and skills and to expand their service remit. For the most part these roles are contained by protocols, policies and PGD's, which, one may contend, confine the specialist nurse, and limit their practice. However, I would purport that these measures also provide safety nets within which to protect our patients, deliver a high standard of care for them and to enable us to develop our practice.

References

- Tye C, Ross F (2000) Blurring boundaries: professional perspectives of the emergency nurse practitioner role in a major accident and emergency department. Journal of Advanced Nursing 31 (5): pp1089-1096
- Nursing and Midwifery Council – www.nmc-uk.org
- Hewitt J (2002) A critical review of the arguments debating the role of the nurse advocate – Journal of Advanced Nursing 37 (5) pp439-445
- Woodward VA, Webb C & Prowse (2006) Nurse consultants: organizational influences on role achievement – Journal of Clinical Nursing 15 272-280
- European Working Directive: www.dh.gov.gov.uk/en/policyandguidance/humanresources

Challenging Assumptions in Ophthalmic Nursing:
a patient centred approach

Professional Practice

What happens when you can't give the level of care you want to give to your client who has a visual impairment, because you have to meet a target imposed by someone who not only is not an ophthalmic nurse but is not even anything to do with health care delivery? Does this become a professional issue for you? Does it affect your professional practice? Professional issues can affect any aspect of the care you deliver. Many nurses these days bemoan the fact that they can't deliver the level of care they want to because of this reason or that reason. So what are they doing about it?

This is sometimes where tensions occur. We are asked to do things or things happen which could (and do) compromise our professional standards. Sometimes we are put under pressure to ask others to act in ways that could be considered unprofessional. This naturally leads to a whole raft of questions.

What does professional practice mean to you? How would you define it to the Martian who keeps popping up from time to time? Do you think that you act professionally at all times when you are working? We are all professionals and we have a responsibility to carry out our care professionally. What is problematic is actually defining what we mean when we talk about professionalism as a concept. Standards of Proficiency (SOP's) - which are referred to in the chapter about

Policies and Procedures – are presented by various Professional Bodies and form part of the education and training of future professionals. These inevitably refer to Professional Practice. Each group of professionals has SOP's which appear to be common across most, if not all, professional groups. Communication, health and safety and equality and diversity are the sorts of issues, which tend to be generic. However each group of professionals will also have 'profession specific' SOP's which relate directly to their particular form of care delivery. What nurses need to be doing is ensuring they are adhering to their particular SOP's to ensure they are maintaining a level or standard of professional practice. The other aspect that we as nurses are bound by is our Code of Professional Conduct (NMC 2002). The Code of Professional Conduct clearly sets out how we as nurses should be conducting ourselves and our practice, the aim being the protection and well being of the patient. However many nurses simply do not use it to support them when they are being asked to act unprofessionally. When I say unprofessionally I mean that we are often asked to compromise our ideals and professional conduct because it simply costs too much. In an increasingly cash-strapped NHS, and in many cases private health care is experiencing similar cash flow pressures, nurses come very close to the wire in some of their acts and omissions. You only have to speak to someone in the supermarket and they are all too pleased to regale you with horror stories about poor nursing practices inflicted either on themselves or someone they know. Of course we defend our nursing and other professional colleagues, but there have been times when it is difficult to justify and defend the practices that you hear about. Whilst we know that people (including ourselves) like to embellish the story and make it often more gruesome than it was in reality, there is still a kernel of truth in the tales told and that is the worrying aspect.

The Code of Conduct is a very powerful tool but is not used to its full potential. When did you last look at yours? Have you got an up to date one from which you can quote chapter and verse? Anecdotally I have found that nurses are ambivalent towards the Code of Conduct and when pressed it's usually because they do not know fully what is contained in it or how it should be used to promote their practice. Many nurses do not understand the concept of vicarious liability. As a result they do not realise that they jeopardise their professional accountability by failing to understand their responsibilities under the Code of Practice. Nurses are becoming stronger and more forceful but there is still a large element of the profession who will put up and shut up. That is, they will just get their head down and get on with it so as not to cause a fuss. Maybe job insecurities play a part in this strategy and nurses feel that if they stand up and are counted then the axe may well fall in their direction. Another worrying aspect is that nurses don't want to raise their heads above the parapet because it is an uncomfortable position to be in. In my opinion this is completely the wrong tack to take. Nurses need to be doing exactly that and standing up for their patients and patient care. Union conferences are good for debating the issues but it is only a very tiny proportion of nurses who attend these kinds of events. When did you last go to an Ophthalmic Nursing Conference? Also it is usually the ones who are not afraid to be counted who will be shouting the odds and heckling the politicians at these conferences. This needs to be extrapolated into every day nursing activities and we need to become more political in support of our patient care. How many of you have lobbied politicians over aspects of patient care which give you cause for concern? How many of you have contacted your local MP to express concern of issues of health care delivery? How many of you have acted as advocates in the interests of your patients? How many of you are members of the Ophthalmic Nursing Forum?

However, political does not always sit comfortably with professional in the eyes of some. Not that long ago it was thought to be most unprofessional to take union action over pay issues. Nurses did it for the love of it didn't they? So why should they want a decent living wage? Nurses too often mutter in groups but won't fight the good fight with all of their might to steal a cliché. Nurses form one of the biggest groups of workers in health care and as such occupy a strong and commanding position. Ophthalmic nurses form a small part of this larger group but usually work in specialised ophthalmic units so are ideally placed to be active on behalf of their specialised patient group. But do we use our strength in numbers? The short answer is no! On the other hand another large group of professionals does very well for itself. Doctors seem to have managed to maintain professional standards by joining forces and offering a united front. Yes of course there are historical factors at play here but if they can do it, why can't we?

Before you accuse me of being a revolutionary (a name I would be more than happy for you to call me though I must say, if it improves patient care), think about what our primary purpose is. We are here for the patient in whatever guise that individual adopts and in our particular case the patient with ophthalmic problems. Patients need us otherwise they wouldn't be entering our hallowed portals. Paradoxically we need patients. If we didn't we wouldn't have very much to do as nurses.

Part of the problem is about the potential to abuse the power and authority that we have as professionals over the unsuspecting public. We do indeed hold very powerful positions over patients. This though has eroded somewhat with the increasingly litigious society in which we live. Patients are more likely to sue us for actual or even fictional malpractice and are actively encouraged by some often less than scrupulous companies to do so. Our professional arrogance has suffered a blow. But that is surely no bad thing. As I keep saying, our prime purpose is to provide care for vulnerable individuals. I remember a discussion with a staff nurse on an ophthalmic ward who informed me she had just taken a phone call for an admission. She told me in no uncertain terms that they had 'a real grot' coming in. It transpired on questioning that the patient had a complex set of needs in addition to her ophthalmic problem, which would require a large input from the nursing staff. As she huffed and puffed over the extra work she and her colleagues would have with this admission I became increasingly concerned over her attitude. Eventually I told her it was my grandmother coming in. Revenge indeed as she was completely horrified. I like to think that maybe she stops and thinks before making such utterances now. However this story is not an isolated one and it seems to me that some of our so-called professional colleagues feel that patients are a necessary evil. This does lead me to question their motives. All of you out there can probably relate similar incidents and anecdotes that in themselves are frankly scary. These stories should be the exception not the norm. You need to look into your own hearts to determine whether you have been guilty of such acts and I leave it to your consciences if you have.

Litigation is a powerful incentive to do the right thing but is it the right incentive? Surely our incentive should be our ability to be professional in all aspects of our nursing lives and by not allowing others around us to either act unprofessionally or to force us into acting unprofessionally. It is very easy to take the line of least resistance and constant battling can be very wearing, which is where of course those that make us battle win! Like I said before I'm not trying to induce revolution but I do want nurses to look at what they do to ensure they are being as professional as they can possibly be. This also gives out a powerful message that nurses will be a hard act to push around if our colleagues can see that we are a force to be reckoned with. Hopefully by acting in a professional manner we can have a greater influence on care delivery because the powers that be will have to listen to us.

This brings us almost full circle to the beginning of the chapter when I posed the issue of defining professionalism as a concept. SOP's go some way towards this but is there a more ethereal connotation to the concept? Is there an element of the gut feeling to consider? Is it something that is instinctive, or is it something learned at a senior colleague's knee? Role modelling can be a very strong influence on us both from a positive perspective and a negative one. We must have all been in the company of senior people we revere and want to emulate. Likewise we have been in the presence of senior people we hope most fervently we are not like. That does not mean that they are unprofessional in their approach. Indeed one person I would never like to emulate was probably one of the most professional people I know! An interesting conundrum, but a powerful one. We can self perpetuate unprofessional behaviour as we are easily influence by those we assume know better than ourselves because they've been around for a good while. Then we end up with a self-fulfilling prophecy. We act unprofessionally so we become unprofessional and so we come full circle.

Nurses sometimes feel that by taking action we actually make things worse for patient care. I had a discussion recently with a very experienced Sister of an ophthalmic ward. She told me that they sometimes had problems getting their patients to theatre, as there is a steep slope to negotiate. Due to the age and physical limitations of many of the ophthalmic patients in her care, wheelchairs and occasionally trolleys are used to transport these people to theatre. However negotiating the slope is causing nurses to complain of back problems from the strain of pushing or from preventing the chair/trolley running away on the downward slope. She told me they couldn't refuse to use this method, as it was the only way into theatre and the patients would suffer if they did refuse because their surgery would be delayed. This would mean that nurses would be accused of acting unprofessionally and indeed when they had voiced concerns had had that accusation levelled at them. I was for once in my life completely speechless. This Sister was being very misguided in what she saw as her professional responsibilities both to the patient and her nursing colleagues. If I was an ophthalmic patient with a post operative patch I'm sure I wouldn't be too happy at the risk of the chair or trolley I was on running away down the slope following my procedure. I wouldn't much like to be the nurse acting as the brake either. No wonder nurses were complaining of back pain. Apart form

Challenging Assumptions in Ophthalmic Nursing:
a patient centred approach

the health and safety issues, by continuing with her head in the sand, in my opinion she was acting unprofessionally to both her patients and to her nursing colleagues. A united front would soon invoke some action (as well as invoking various health and safety legislation) and managers would have to sit up and listen. However, in my experience nurses are very good at holding each others coats whilst the one who is not afraid stands up to be counted and has to beard the lion (manager) in their cage alone. This dilutes the message making it much less powerful. It also singles out the one who is not afraid to stand up and be counted. Whereas if we stood shoulder to shoulder (think suffragette here) we could show a united and professional front to deal with important issues.

Unfortunately very often professional issues are very much enmeshed with the wider political picture that faces most of us these days. Whether we like it or not we have to be politically aware to enable us to uphold our professional principles. For many nurses this is not a comfortable position to be in. They want to deliver the best possible care but in many cases want to do this almost in a cocoon without wanting to get involved in the wider issues and implications. That may have been fine in the 'olden days', but cannot possibly work in today's many nursing arenas. We cannot and should not just shut up and put up. We have to fight our corner whatever that corner might be. So you ophthalmic nurses out there reflect upon your practices, reflect upon the stance you take, reflect upon how you can uphold your professional principles. It is down to you and your ophthalmic nursing colleagues to insist upon acting in a professional manner at all times, even if that means that you come to blows with those in managerial positions. Remember though that you must act in the best interests of your patients. Don't just use this as an excuse to get back at managers who, after all, have their own set of complex hassles to deal with. Another good approach to take is not to present the 'problem' to your manager, but tell them what the problem is and what can be done to solve it!

I would strongly suggest that you take time to revisit your Code of Professional Conduct to make sure you are understand it and are upholding it.

References

- Nursing & Midwifery Council (2002) The Code of Professional Conduct. NMC. London

Challenging Assumptions in Ophthalmic Nursing: a patient centred approach

Policies, Procedures, Standards and Audit

Boring or what? Four topics guaranteed to make anyone nod off…until now that is. I want to show you how all of these both individually and collectively can be used to your advantage but even better to your patient's advantage. However, it's not all plain sailing by any stretch of the imagination and there are a lot of questions that need to be asked and issues to be addressed before we have a refined and usable product. Whilst all of these issues are inexorably linked and intertwined I want to attempt to separate them out for the purpose of discussion and then reassemble them.

Policies and Procedures

These are always spoken of in the same breath, but are they the same thing? Does one fall naturally from the other? Indeed can you have one without the other? Could you differentiate between the two and explain it to any friendly Martians who may be visiting? This for me is the acid test. If I can't explain it do I really understand it? If that applies to you perhaps you need to take some time out and consider what you really understand by these two terms.

Generally the policy will tell you what needs to be in place and the procedure will tell you how it should be done. However some policies perform both functions by telling you what action you need to take in the event of something happening. Manual Handling policies very often perform this dual function. Policy, by virtue of its name, implies some aspect of control or policing whereas the procedure is simply a guideline on how to do something. One of the issues (I don't like to call them problems) with policies is that there seems to be so many of them. This inevitably leads to conflict. One policy will say one thing whereas another one will say the complete opposite. A legislative example might be the Data Protection Act (1998) and the Freedom of Information Act (2000). In your workplace you may find that there are similar conflicts with policies and procedures. Have a look through your policy file and see if you can find any. Another issue with policies is that very often they are used as "a blanket". This is not necessarily the fault of the policy but rather a knee jerk reaction to something that has happened. In these days of litigation for the slightest of reasons (or excuses in some cases) the response very often is to put a blanket policy in place as a cure all. This has the effect of holding back innovation, as people are afraid to step outside of the boundaries of the policy. Blanket policies can also be directly detrimental to patient care. How? I hear you cry. Well I'll tell you.

The Human Rights Act (1998) is a good thing isn't it? Or is it? Problems arise because both the patient and the carer have the same human rights. Shadow cabinets and some within government have already questioned the concept. It may well be abolished by future governments and replaced with something more workable. Inevitably there is going to be conflict. A rather unwieldy example is given below. The policy is that you will use universal precautions come what may and you will never deviate from this. You as the nurse may require patient X to undergo a simple blood test, which you are competent to take. Patient X however does not want you to wear latex gloves as he feels this is demeaning and implies he is dirty or has some foul disease. He feels it is infringing his right to be treated with dignity. You explain that it is for general protection as it is not good practice to share body fluids unless you are married or in a long-term relationship! You also have the right to be protected from harm (in this case blood borne disease). But he is insistent. Who is right? Who wins? Both are right and no one wins. You have the right to protect yourself; he has the right to be treated with dignity.

At the policy level one could argue that rubber gloves give us a false sense of security. Some believe that they provide an impenetrable barrier between us and those nasty bugs that people carry and try to inflict upon human kind. I did once work with one fairly senior nurse manager who made her nurses wear two pairs of gloves just to make sure. When I pointed out to her that a needle will go through two pairs as easily as it will go through one pair, she realised that she was being over cautious. It also appeared that the nurses were only removing the top pair of gloves once a procedure was finished and applying a new pair on top for further treatments. Not a hand-wash or alcohol-rub in sight. Very hygienic!

Another argument might be that the gloves can make practice sloppy for the very reasons the nurse manager was attempting to protect against. Because we wear gloves we are not as careful about our use of needles and the like. Now I am not saying that we shouldn't undertake universal precautions, indeed they are there for very good reasons but likewise we should never say never. This is where blanket policies cause problems. They don't allow us the freedom to never say never. In a first aid situation would you say that you couldn't deal with someone who was bleeding profusely unless you were wearing your gloves? Then what happens when your manager spots you not wearing your gloves, do you get involved in disciplinary action even though you were performing life-saving activities?

Blanket policies are in abundance. What we very often don't realise is how many of the policies are blanket ones and how they can prevent innovation and freedom of professional thought. There have been judicial rulings on blanket bans involving no lifting policies (Disability Rights Commission 2003). In this case, the judge ruled that carers could not ever say that they would never lift a client, which is a direct result of the judge applying the Human Rights Act (1998) in favour of the client. As has been stated before, policies sometimes come about as a result of an incident happening. However there often seems to be a lack of joined up thinking. Rather than looking at what has gone on before, someone, usually a manager although this is not always the case, brings in another policy. This may or may not take account of the previous policy and indeed it may be an add-on to an existing policy. Then before we know it we are so tied up with policies that we can no longer exercise our professional autonomy. Rather than just adding policies we need a systematic review of policies to ensure we operate safely but are not constrained. I have direct experience of this scenario whereby a hospital I worked for would not allow trained agency nurses to hold 'the keys'. This was because an agency nurse had once taken some ampoules of a controlled drug. This blanket policy prevented trained agency nurses from exercising their professional autonomy and accountability. This even applied to nurses who worked for the hospital but did some agency work for extra cash. Crazy or what? The implication of course was that agency nurses were not trustworthy on top of which the potential impact on a ward if the trained nurse can't 'hold' the keys.

Procedures

Procedures however are a different kettle of fish. These are about saying how things should be done. They very often present a step-by-step process which one has to go through to achieve the aim of the procedure. Whilst guidelines can provide a useful framework for the delivery of care sometimes they can be too prescriptive. Have a look at your own ophthalmic procedures. How many of them adopt the 'Noddy and Big Ears' approach and treat you like a complete imbecile incapable of rational, accountable and professional thought?

Other things to note are:
- When was it written? Is the paper similar to parchment in colour and consistency? This is usually a dead give away that it was written sometime around the dawning of civilisation.
- When was it last reviewed? This relates to the previous bullet point.
- Who actually wrote it? Was it the clinicians responsible for your patient's care or was it some faceless office-bound individual who hasn't got the first idea about delivering ophthalmic care?
- Where is it kept? Do you know where it is or what it says? Could you explain it to that Martian who cropped up earlier?
- How well used is it? Pristine paper with no discernible marks is very suspect as it implies that no one has touched it let alone looked at it.
- Does anyone monitor its use? Has it ever been audited?
- Are there policies in there that you didn't know existed and are therefore in breach of?
- Are all of the policies in your department necessary?
- How many of them are actually evidence-based? How would you know?

There is also the question as to whether we need all these local policies and procedures. Some places simply use published ones as a way of not reinventing the wheel. Renowned centres of excellence have usually published these and it would seem to make sense to utilise the ones that pertain to your particular area. They are generally tried and tested and are evidence-based. They are also regularly reviewed and updated and republished.

Standards

Standards set the level of attainment required. They are part and parcel of healthcare today and are constantly being measured either by yourself, your manager, your manager's manager or by the audit department (more about that later). If you have anything to do with National Vocational Qualifications (NVQ) you will be familiar with care standards as these are the basis of the NVQ framework. Standards are an integral part of the care we deliver but how many of us know what the standards are that we are being measured against? Can you, off the top of your head, identify particular standards associated with particular aspects of your care delivery? There are of course the National Service Frameworks (NSF) but is there one specifically for your area of care delivery? Of course ophthalmic patients don't always just have an ophthalmic problem they very often have other underlying health problems so may come into one of the NSF's but this is not necessarily for their ophthalmic problem. Other standards you may have thought of include the Knowledge and Skills Framework (KSF) (DOH 2004) that underpin Agenda for Change (DOH 2004). These stipulate the standard you need to be able to demonstrate to show where on the pay scale you can sit. This has most certainly concentrated the minds of most workers in the NHS.

There are also shed loads of hidden standards that you may only be aware of subconsciously. You know they are there and you know you should be working towards achieving them. These will include such things as the standards for hand washing; for dressing techniques; for drug administration to name just a very few. These standards may be an integral part of the relevant policy or may be something completely separate. However they may not always be written down somewhere so how do you go about measuring achievement of the standard? When people complain about an aspect of their care experience, how do you determine whether a standard has been achieved or not and which standard it is that has been breached? I have heard people say that the standard of care is not good in such a place. My question is what is that standard of care? Whose standard is it? Yours? Theirs? Mine? Does the standard come up to the expectations of the relevant stakeholder? Do relatives expect higher (or sometimes lower) standards than the actual standard states? Are we all working to the same standards? Some people allegedly have higher standards then others. How does that work then?

Think about your own working practices. How do you know you have achieved 'the standard'? It might be useful at this point to

actually write down what measures you use to assess the standard of care you have delivered.

Many of the governing bodies for health professionals (Nursing and Midwifery Council; Health Professions Council to name just two) use Standards of Proficiency (SOP's) (NMC 2005, HPC 2007) as the basis for their care delivery. These SOP's form the basis for the professional training and education of the relevant health professional and provide a useful framework of standards by which an individual can be measured. They are profession specific, although there are obvious commonalities such as communication and equality and rights. These go some way to enabling professionals to be measured for their particular aspect of care. We the professional can also wave them at the unsuspecting public to show that we use standards when we look after them. A colleague of mine who worked in the community was very unsettled by a relative who always kept a copy of the Guide to Handling of Patients (Back Care 1998) with her when my colleague visited. She would look up items to do with moving and handling as my colleague moved or handled her loved one. My colleague never did find out whether she was checking up on her or whether she was just interested in what was being done to her nearest and dearest. It is interesting though, that she was unsettled by this experience because if she was working to the standard she would have nothing to worry about. You may have had similar experiences if you've ever had to care for someone with even a smidgen of ophthalmic knowledge. I am currently being treated for a relatively common ophthalmic condition and my ophthalmic consultant certainly treats me with a certain amount of caution. So are we a bit on the defensive if anyone questions our practice for this very reason? You will have to talk to your own conscience about that but whatever the outcome of your discussions with yourself standards are a measurable entity, which can demonstrate quality. They can also be used as part of performance management to determine how well you are doing and whether any corrective measures need to be introduced.

Audit

Audit is a term that is bandied around by people in the know on quality matters. It is a way of measuring quality, whatever quality is. Audit merits a book in its own right so I am not going to do more than scratch the surface here. What you need to be thinking about though is your participation in the audit process. We tend to be passive players when it comes to audit and it is one of those words that is used as a threat when things are perhaps not going so well. However audit is a very useful way of opening the eyes of the people who are part of care delivery and that includes everyone from the lowest of the low to the highest of the high. No one should be immune to audit. Take your own working area. Is there anything that immediately springs to mind that could do with an audit? You may need to have a bit of a brainstorm with your work colleagues to come up with some ideas, or you may well have ideas of your own but have not yet got round to doing anything with them. Audit is not just about care delivery but can be about the furniture and fixings, as all of these will ultimately have an effect on the care which we deliver. Simple things can warrant audit whereas very often we try to look too deeply. Think about part of your ward or department. How accessible is it for your ophthalmic patient? Think about all of the different issues that affect the ophthalmic patient. Remember they may well have other problems in addition to or on top of their ophthalmic problem. Don't just think about their ability to see (or not see as the case may be). If they are elderly they may have mobility problems; if they are disabled they may have general accessibility problems. But because we work in the setting day in and day out we don't always see it. Sometimes we need to step back to take an objective view but don't completely disregard the subjective view as this can be equally important. Try a bit of empathy. Imagine you are one of your patients (pick any patient you like) and walk around the department and see it from their perspective. May be you could even borrow the training glasses and really get the feel for your ophthalmic patients.

Once you've decided on what you are going to audit you will need to refine the idea and to narrow the focus of your investigations. If the focus is too broad it may well become unmanageable. You will also need to get as many people on your side as possible so keep people informed of what you are doing every step of the way. People feel threatened if they think that they or an aspect of what they do is under investigation. We as a species are very suspicious of people's motives so have no secrets or hidden agendas. Remember audit can show what you are doing well, as well as showing up any deficits in any aspect of care delivery or the environment in which it is delivered. Always publish (in the broadest sense of the word) your results. Be open and honest in what you find. Improvements can only be made if relevant individuals know what those improvements might be.

A word of caution, Audit needs to be properly managed. It is not just about ticking boxes. It should be a thought out process, which is properly planned and communication must be open and honest. It is a process well worth going through as it enables you to see things that you may not have seen or thought about before. You will be part of a drive for quality improvements, which not only should improve the patient's lot but may improve yours as well. So next time someone mentions the word audit, don't cringe…applaud.

Conclusion

Policies, procedures and audit can all be made to work for you and this will ultimately make improvements to the experiences of your patients. Policies and procedures are there to protect both you and your patient but can become unwieldy if not managed properly. You are in an ideal position to challenge managers when a new policy is

Challenging Assumptions in Ophthalmic Nursing:
a patient centred approach

introduced. You can explore why it has been introduced and whether it complements or constrains existing policies. Don't be afraid to ask questions. Think of it from the perspective of your ophthalmic patients and ask from the position of advocate for their care.

Audit is about measuring how well (or poorly) a standard has been met and should be used to build on strengths but also to correct weaknesses in the system. Don't fear audit but relish the issues and challenges it raises for you in the care of your ophthalmic patients. If there is an audit department in your place of work, go and see them and discuss issues you feel might warrant auditing.

References

- Back Care (1998) The Guide to Handling of Patients 4th Edition
- Department of Health (2004) Agenda for Change. DOH. London
- Department of Health (2004) The NHS KSF and the Development Review Process. DOH. London
- Disability Rights Commission (2003) www.drc.org.uk
- Health Professions Council (2007) Standards of Proficiency. HPC London
- HMSO (1998) The Data Protection Act. HMSO. London
- HMSO (2000) The Freedom of Information Act. HMSO. London
- HMSO (1998) The Human Rights Act. HMSO. London
- Nursing & Midwifery Council (2005) Standards of Proficiency for Pre-registration Nursing. NMC London

Challenging Assumptions in Ophthalmic Nursing:
a patient centred approach

Models of Nursing in Ophthalmology

So, are models of nursing a help or a hindrance? Well that will depend upon your point of view and probably your experience of them. As with a lot of influences in nursing and health care, nursing models originated in the USA. As a student nurse in the late 1970's I was introduced to the delights of Virginia Henderson (1968) and later the British version from Roper, Logan and Tierney (1980). Both of these models were simple concepts and were a bid to move away from the medical model approach to care delivery and were part of and very often underpinned 'the nursing process'. The whole idea seemed to take off and began to grow almost exponentially. Later models were developed and became increasingly complex and in some cases downright obscure! There seemed to be a move away from simple concepts with each emerging model vying with the previous one to be more conceptual and complex. The cynical amongst you may well question the validity of nursing models. Indeed were models designed to be used in clinical practice or were they simply a theoretical concept that authors devised as part of a thesis for a first or higher degree? Whatever the answer nursing models are here and are part and parcel of nursing today. As a matter of fact they are looked for often as a measure of quality of care by the various bodies such as the universities, which inspect clinical environments, as part of an audit process for the various learners supported by these clinical placements. Whilst this is the received wisdom and one can see why they can be of use, we need to challenge the assumption that they are the be all and end all in the quest to deliver first class nursing care. We need to be actively interrogating their applicability to and enhancement of ophthalmic clinical nursing practice.

Advantages of nursing models

Nursing models allow professional nurses to make nursing decisions based on a nursing ideology rather than a medical one. A model enables holistic care to be delivered on an individualised basis to the people in our care. A model can be tailored to a particular speciality so that it fits in with the general ethos of the area and doesn't contain needless information. As the medical model provides a framework for doctors, so then do nursing models provide a framework for nurses? They can act as an aide mémoire so that the nursing staff can be sure that all the relevant information pertaining to the patient is collected in a structured way. A good model will allow you the ophthalmic nurse to build up a picture of the patient as a person and individual. It will usually contain not only nursing specific information but also social and medical details. From this a detailed care plan can be developed to ensure the patient receives care appropriate to their identified needs whether directly related to their ophthalmic condition or as a result of something else. These interventions can then be individualised so that the patient is on the receiving end of a tailor made package. It will also enable evaluation of nursing interventions so that changes and improvements can be made to the care delivery as the patient progresses through our ophthalmic care systems. This makes the 'nursing process' dynamic in the ever-shifting sands of care. The old adage that quality is a journey not a destination fits nicely into this philosophy. Ophthalmic care delivery should change as the needs of the patients change, hopefully with recovery but also with deterioration. We should be learning from our evaluations and the good model will enable and facilitate this process. In short the nursing model will provide a wealth of information to enable a patient to receive the optimum ophthalmic nursing care.

In this context, which may (or may not) match reality, nursing models are obviously the saviour of the modern nurse! Nonetheless as with most things in real life there is another side to the argument and there are disadvantages to nursing models in ophthalmology.

Disadvantages of nursing models

A nursing model is only as good as the person obtaining and recording the information. This in turn is dependent upon the model used in the first place. As stated in the introduction, as nursing models developed so did their complexity. It is possibly for this reason that the less complicated ones seem to have endured. Never the less, if the person completing the assessment does not really understand the underlying concepts of the nursing model, the appropriate information may not be recorded or is recorded incorrectly. Additionally if the model is complicated, the individual who is obtaining the information and completing the documentation may pay lip service to the information requirements and omit what to the carer seem to be needless questions. Added to this are time resource issues. Nursing documentation needs a substantial amount of time to complete to a suitable and sufficient standard. This may be fine if the ophthalmic area you are working in is an environment that may have a slow through-put but on a busy acute unit that time may not be available to you. Additionally the model may be constraining, restricting innovation as nurses adhere to the received wisdom that is the nursing model. Nurses are then afraid to step outside of the remit of the model for fear of retribution by the powers that be whether managerial or statutory. This stymies nursing intuition, preventing ophthalmic nurses from looking outside the box and maybe working with their professional instincts. This is even more apparent with the introduction of pre-printed, prescribed documentation that nurses feel they cannot detract from and must stick to rigidly. However this is not necessarily the fault of the model. Perhaps the blame for this should be laid at the feet of those who manage the care processes who have determined that it is more economical in terms of time for the documentation to be 'off the peg' as opposed to a tailor made approach.

Consider for example those outside of the immediate care process, but who also are directly affected by what happens during and after that care process. What of the mother from a previous chapter who has been told that her young son not only is about to go blind but also has a potentially life threatening illness. Where do they feature in the nursing model?

Discussion

It would appear from the dialogue above that there are more disadvantages to nursing models than advantages. This is perhaps an unfair assertion. As discussed above, models are usually only as good as the person using it. The reality may well be quite different from the original concept devised by its author. Maybe the concepts are just that – concepts – devised to wake up the nursing profession and make nurses think about what they are doing in terms of delivering care. It also promotes thinking outside of the box, empowering nurses to be more innovative in their approaches to ophthalmic care delivery. However we need to ask the question as to what the intention of the authors of these models was? Did the authors intend for the model to be utilised wholesale as it was written or was it just a way of getting nurses to be more structured in their thinking? Without interrogating the authors it is probably not possible to answer this satisfactorily and we can only conjecture. Whatever the answer, we do need to question models of nursing and their applicability to our field of practice.

However, having such a plethora of models can be confusing for the clinical practitioner especially when trying to choose and implement one that is meaningful and workable. Yes it is good to have choice but if we are to look at other professional groups they don't seem to be so hung up on bodies of knowledge generally, and models of their own kind of care delivery specifically. There are so many nursing models jumping out of the nursing literature they are almost diluted and it certainly makes choosing an appropriate model for ophthalmic nursing a difficult task. It's like the academic branch of nursing is in competition with itself to produce these theories and associated models without seriously considering the impact on the poor souls who have to implement them.

Where the Allied Health Professionals (and I include doctors in this group) score highly is in their inherent belief in what they do. They use a systematic approach to their particular care interventions but don't seem to need to justify their use of it. They just get on with it.

There is certainly a need for a systematic approach (and there are enough of them in nursing) but why so many? One basic template, similar to the methodology of other professions, which can be personalised for particular areas or specialities within nursing, might be a better approach. Better still, why don't we try to incorporate aspects of our fellow professionals' approaches which will facilitate inter-professional working? This is very much on the agenda for care and there is a great push towards multi-professional and inter-professional learning and working. A universal approach would greatly assist in this and in the education of future professionals. Many ophthalmic units already have good inter-professional links but each AHP has its own framework that potentially impacts on patient care as each group does its own thing and records it in its own place.

However, with this notion there is an immediate issue that raises its ugly head. Which systematic approach should we adopt? Which model is the best one to use? How does it fit in with the inter-professional agenda? This is where you the practising ophthalmic nurse come in.

Models in Practice

Try this very simple exercise.

Have a look at your area of practice.

- Do you use a model of nursing?
- If so (and I bet you do) which one is it?
- How well do you know it? Could you describe it to a Martian so that they would understand it and be able to use it?
- Who said you had to use it?
- What input did the clinical nursing staff have into its use?
- Did you have any training in its use?
- How does it tie in with your nursing documentation? Is your nursing documentation built around the model or is the model

built around the documentation?
- How effective is it at assisting you to deliver care?
- Has it been updated recently?
- Have other models been tested at all?
- Do you actually in reality use it or do you pay lip service to it? (Come on now be very honest with yourself, you don't need to tell anyone yet)
- What would the consequences be if you didn't use it? (Think broad as well as long)

Other things you may wish to consider are:
- How appropriate is your model to ophthalmic nursing?
- How does it fit in with the other areas within the ophthalmic nursing fraternity? Is it the same or different from that used in other ophthalmic departments?
- Does it miss anything out which is specific to ophthalmic nursing?
- Does it contain information that is not relevant to ophthalmic nursing?

You may have never questioned or even thought about any of these issues before now, but I hope you are now starting to take a step back and view them objectively. It is easy when working with something just to accept it 'warts and all', usually because it has always been there and it will do. There is also the notion of 'if it 'aint broke, don't fix it' but this can be equally restrictive and prevent innovation in nursing practice.

Another simple exercise you can try is to design your own model of ophthalmic nursing. Ignore what you already have and start with a blank piece of paper. Now make a list of all of the things that you consider need to be included in any nursing documentation to enable you to deliver optimum ophthalmic care effectively and efficiently. Don't think about existing documentation but be as innovative as you dare. It's what you consider to be important that counts. After all you are the one who is going to be using it.

When I've got ophthalmic nurses to design a model in a classroom setting where they can be meticulous and take time to get the creative juices flowing, very often they include many aspects of the medical approach to care delivery. This clearly leads to the question why? One could argue it is because nurses are indoctrinated into a medical way of thinking early on in their nursing experiences and this is perpetuated throughout their career. This in turn is passed on through to the students who are exposed to this way of thinking and delivering care. If nurses are not committed to nursing models or don't really understand them and their place in nursing then this situation will never be resolved. Certainly qualified nurses who I come into contact with generally are unimpressed with models, paying lip service to the concept. Perhaps this is because it has never been explained to those who use it and those who use it have never questioned the reason or need for its use. It may also be about comfort zones. Nursing is high pressure these days and it is easier to get on with it, head down and don't ask questions. The worry being that if you do question anything you are either labelled as a troublemaker or you are then given the task to sort out!

However, one may argue that the medical model is tried and tested and works, forming a useful basis for the development of nursing models. Rather than chucking out medical ideology nurses can see that there is a time and place for them. However this is not to say that the medical model is the best. It does have downsides, predominantly that it is not based on nursing care delivery.

A major issue with nursing models is the fact that they were often imposed on the workforce by 'management'. This then means that those who have to use the model do not have ownership of it and tend not to wholly embrace it. It is well known that participation increases motivation and if you participate in the development of the model you will be more likely to use it. The model is often regarded as an encumbrance rather than aiding the busy nurse. As time evolves the model is adopted but may not be totally understood and so is not always effective in assisting in care delivery. Straw polls with nurses over a number of years have shown that nurses generally (and this is a broad generalisation) have very little insight

Challenging Assumptions in Ophthalmic Nursing:
a patient centred approach

into nursing models and their use. Think about a time when you've changed jobs or moved to another area maybe due to reconfiguration. Did you question the model or the documentation and its applicability? Nurses commonly just use the documentation they are given without question regardless of whether it fits in to the ward or department philosophy or mission. Very often is seen as a necessary hindrance or evil; another straw to break the camel's back. You, the professional ophthalmic nurse needs to be examining the use of models of nursing in your own practice and asking questions as to its practicability and applicability to your own branch of ophthalmic nursing. That's not to say that I advocate wholesale abandonment of models, I just ask for you to be more questioning in your use of them.

Why do we have to use Roper et al (1980)? Why can't we develop the ophthalmic model of nursing based on the best bits of all of the other nursing models? Maybe some of you do and have the X or Y nursing model applicable to your area. Sadly this is not always acceptable to the powers that be who want named models. But surely this is more about them not understanding the concept properly. It might be to do with academic snobbery but perhaps dare I say it that the powers that be only recognise published models because they don't really understand them. A published model must be good because it is published, but if you can prove that your particular model fulfils the need for a systematic approach then that should be sufficient. There is a way round this of course. Develop your own model and publish it. Others can then either use it or abuse it but at least it's published. We need to be more proactive in deciding how we deliver our particular brand of care. Nurses are notoriously poor at arguing their case in the face of opposition. Even today, with more and more degree and diploma educated nurses on the circuit, there seems to be reluctance to question or counter-argue issues because of a lack confidence. There seems to be the belief that some one 'better than ourselves' knows best. Often that 'someone' is usually divorced from the practice environment and has forgotten the pressures involved in delivering care. They might be under pressure themselves though to ensure the practice environments tick the audit boxes. We cannot hold them solely to blame though, we must accept some of the responsibility for not speaking out and arguing our corner.

Very often practitioners feel constrained as things are imposed on them from – in their perception – above. This of course may not necessarily be the case but it is certainly how it is perceived. This imposition leads to a lack of ownership by the practising nurse that in turn can lead to apathy when trying to utilise a model of nursing effectively. This apathy seems to prevent practitioners informing the process for whatever reason. This could be the head down approach adopted by some or a lack of comfort with the whole concept. It might also be because we feel that we have to accept what is imposed as it is too late to do anything about it. Cynics will often question the notion of consultation feeling it is just hoops that management have to jump through. What management do with the results of the quasi consultation is moot point. Whatever it is nurses take a lot of convincing that anything meaningful comes from it. Those of us who have been around for a fair amount of time find that if we stand still for long enough we will see what we did before come around again all in the name of progress.

You, the practising ophthalmic professional, need to speak up and out. You need to be in charge of the nursing care that is delivered to your client group. You are the experts in your field. Theorists may have been once but may not be now.

What the theorists have done however is to break the mould of classical, medicine-led nursing and have opened up the channels to produce and provide high-class nurse-led care. Don't just throw out the model but question its validity in your field of ophthalmic practice and if need be write your own. What we do need however is a unified approach that can embrace inter-professional practice that will avoid repetition. That includes those elements of the medical model that work and maybe elements of the assessment tools used by the AHP's. Think of models in a positive light but make them work for you and ultimately your patient.

References

- Henderson V (1968) ICN Basic Principles of Nursing Care. ICN Geneva
- Roper N, Logan WW, and Tierney AJ (1980) The Elements of Nursing. Edinburgh. Churchill Livingstone.

Challenging Assumptions in Ophthalmic Nursing:
a patient centred approach

Conclusion

So there you have it. Having read the book from cover to cover (we hope) or even if you are a dipper, we hope we have challenged you the practitioner or care giver to think about what you do with, at and to your ophthalmic patient. We set out to be deliberately provocative as it is an approach we feel needed to be adopted. There is too much political correctness today and we wanted to turn that round and get to the crux of the matter - the patient. We have many years combined experience in patient care between us and we wanted to share some of our thoughts and the numerous discussions we have had over the years. Our prime concern is and always has been that the patient is losing their identity. They have become a condition or a collection of conditions rather than a human being with a set of needs, which we the nursing staff should be assessing and developing a plan of action for. We wanted to make you sit up and think rather than work as an automaton on overdrive without thought about what you are actually doing.

We hope that some of the things we have written have made you distinctly uncomfortable either on a personal basis or on behalf of people with whom you work. Most of all we hope that reading this book will help you to take stock of your own practice and the practice of others and really give you the incentive to turn things around. Now this is going to be a bit like turning a tanker round at sea, not something which is done quickly. However, our book is here for you to turn to or at the very worst throw at someone!

What about us? Is our job finished now with this book? No, far from it. What started out as a cathartic process for us has now grown and developed. Rather than being the end, since we have now 'got it off our chests', it has moved us onto other things. The final chapter of the book talks about nursing models in ophthalmology and that's where we are off to next. We want to explore and develop this concept to consider the viability of a nursing model specifically designed for ophthalmic patients. We have already started those discussions so watch this space.